Exercises for the Zoology Laboratory

REVISED SECOND EDITION

David G. Smith
University of Richmond

Morton Publishing Company
925 W. Kenyon Avenue, Unit 12
Englewood, Colorado 80110

http://www.morton-pub.com

Book Team

Publisher   Douglas Morton
Design/Production   Ash Street Typecrafters, Inc.
Cover Design   Bob Schram, Bookends

PHOTO CREDITS

Front Cover: (clockwise)
American kestrel . . . . John A. Smallwood
African tree frog . . . . Rafael O. de Sá
Luna moth . . . . W. John Hayden
Ornate box turtle . . . . David G. Smith

Back Cover:
Green darner dragonfly . . . . David G. Smith
Yellow-bellied marmot . . . . Gayl K. Ortiz

Printed in the United States of America

10   9   8   7

ISBN-10: 0-89582-618-6

ISBN-13: 978-089582-618-3

Preface

Exercises for the Zoology Laboratory is designed to provide a broad, one-semester introduction to the field of zoology. A basic understanding of and appreciation for the diversity of animals that exists is fundamental in establishing a secure foundation for more advanced courses in biology. This manual provides a series of observational and investigative exercises that delves into the anatomy, behavior, physiology and ecology of the major invertebrate and vertebrate lineages, thereby giving a broad survey of the animal and protist kingdoms in an evolutionary context. The order of coverage follows a logical progression beginning with animal cells and tissues, then advancing phylogenetically from protists through chordates. The content and breadth of the material is primarily geared toward introductory- or intermediate-level university courses concentrating in zoology.

New to the Second Edition is a chapter on protists. Because of recent discoveries in molecular and cellular biology, this large group of organisms is in a state of taxonomic flux, undergoing extensive reclassification as new evidence surfaces. To keep current with the changing face of taxonomy, this edition contains the latest classification scheme for protists, paralleling many of the leading textbooks in the field.

The format of this manual remains the same and is structured to be used with *A Photographic Atlas for the Zoology Laboratory, 6th Edition*. This edition of *Exercises for the Zoology Laboratory* contains updated page references to the detailed color illustrations and photographs provided in the sibling atlas. While either guide can be used alone with success, the loose leaf design of these books allows them to be combined into a comprehensive manual that paints a more complete picture of zoology. I strongly recommend purchasing the two manuals and using them in tandem for maximum educational benefit.

Great care has been taken to provide information in an engaging, user-friendly manner that both students and instructors will appreciate. Each chapter begins with a list of laboratory objectives and "materials lists" accompany each exercise. Instructions to students are numbered and set off from the main text, while important terms are **boldfaced**. There are questions designed to allow students to monitor their progress through each exercise as well as review questions at the end of each chapter. Numerous illustrations and photographs are provided and tables are used throughout to conveniently summarize information. A glossary containing definitions of all boldfaced terms and an index are provided for quick reference.

Chapter 1, "Fundamental Laboratory Skills," ideally should be read *before* coming to the first laboratory session to provide the necessary groundwork for microscopy, the metric system, making wet mounts, etc. Its main purpose is to level the playing field of basic laboratory skills so that all students enter the course with competency in these crucial areas. Subsequent chapters build upon these important skills.

It is my hope that *Exercises for the Zoology Laboratory* provides instructors and students with an illuminating, hands-on view into the fascinating world of animals. I hope that your journey is both stimulating and rewarding, and I welcome your comments and suggestions for improving this book.

— David G. Smith

Acknowledgments

I am indebted to many individuals who participated in the compilation and review of this material, without whose assistance, diligence and dedication I certainly would not have been able to complete this book. First I would like to thank the many people at Morton Publishing for the opportunity. Joanne Saliger and her team at Ash Street Typecrafters did a fabulous job with the design and layout of the book. Many thanks to Bob Schram at Bookends for the beautiful cover. Jon Glase, Mel Zimmerman and Jerry Waldvogel graciously permitted the use of their ideas on orientation behaviors. Many thanks to Charles Drewes who was overly generous with his time and provided insightful and wonderfully creative ideas for investigations into the behavior and physiology of sponges and blackworms. Darryl C. Smith, M.D., played a vital role in the development of the latter chapters on vertebrates. In addition, he extensively reviewed the manuscript and provided valuable editorial comments and constructive criticisms that polished the presentation of the material and ensured its accuracy. Thanks for the keen eye, Darryl. Finally, I would like to thank Terri Cain for her continued support throughout this project and for her assistance in assembling the final manuscript.

Contents

Laboratory Safety

Safety guidelines in biology laboratories are often taken with a tongue-in-cheek approach. Many students casually think, "What could happen to me in Bio Lab?" While you certainly don't often run the risk of catching yourself on fire or blowing up something (risks more akin to a Chemistry Lab), there are plenty of avenues for accidents of all sizes in the biology laboratory. Integral to learning about biology is accepting the responsibility that comes with doing biology. In your pursuit of knowledge about the animal kingdom you will immerse yourself in a multifaceted assemblage of observations, dissections and experiments, each with their own specific protocols, techniques and inherent perils. Do not become complacent about the purported sanctuary of the biology laboratory. Accidents can happen and will happen as soon as you let down your guard. A common complaint is that too many rules take the fun out of lab. Common sense dictates that some degree of "looseness" must be sacrificed to gain the necessary degree of safety that will ensure that your laboratory experience is positive. The laboratory is no place for a carefree, haphazard attitude. However, in the proper perspective, these basic guidelines will keep you safer and happier and will teach you the appropriate protocols that allow biologists to study their organisms safely and effectively.

The following list of basic safety rules for the laboratory is offered as a guide to make your laboratory experiences safe and enjoyable. It is by no means a complete list, but rather a starting point upon which you can build a tailored list to suite your specific laboratory. Remember, your best defense against accidents in the lab and your greatest asset in dealing with situations when they arise is *common sense*. (When *that* fails you, alert your instructor!)

Basic Laboratory Safety Guidelines

- Never eat, drink or smoke in the laboratory.
- Keep your hands away from your mouth, eyes and nose as much as possible during lab.
- Wear close-toed shoes that adequately protect your feet.
- Wear protective gloves and/or goggles when handling or dissecting preserved specimens.
- Keep track of the materials on your workbench and keep your workbench uncluttered.
- Place any disposable or broken glass in marked containers designated "Broken Glass."
- Place disposable scalpel blades or other sharp metal objects in their properly designated containers.

- Replace dull or broken scalpel blades. More accidents occur with dull scalpels since more force is needed to cut with them, increasing the chance of slipping.
- Know the locations of the first aid kit and eyewash fountain in your laboratory and know how to use them. Your instructor should discuss these safety items during the first laboratory period. Ask about them if they are not discussed.
- Report any electrical anomalies to your instructor immediately (e.g., frayed electrical cord, bare wires, broken plug, foreign object in socket, faulty switch).
- Alert your instructor in the event of an accident—no matter how harmless it seems! There may be unseen dangers of which you are not aware.
- Any contact with human blood should be reported to your instructor immediately.
- Clean your lab bench and other work surfaces at the end of each lab period.
- Wash your hands carefully with soap and warm water and rinse them thoroughly before leaving the laboratory.

Fundamental Laboratory Skills

After completing the exercises in this chapter, you should:

1. Understand metric weights and measurements.
2. Understand the basics of pipetting solutions.
3. Be familiar with a compound microscope and its use.
4. Be able to make a wet mount of a specimen on a microscope slide.
5. Be familiar with basic dissection techniques.
6. Understand references to body symmetry, body planes and body regions of animals.
7. Be able to define all boldface terms.

Metric Weights and Measurements

> **Materials needed:**
> - meter sticks
> - metric rulers
> - staples
> - graduated cylinders (50 mL, 100 mL, 1000 mL)
> - coffee cups
> - gallon jugs
> - paper clips
> - electronic or triple beam balances

Scientists throughout the world use a standardized system of weights and measurements, the metric system. This system has been adopted by virtually every major country around the world with the exception of the United States. As a result, American students are often not as familiar with the relationships of metric units. Yet the metric system has crept into our society in a few areas (2 liter soda bottles, 35 mm film, 9 mm handguns, etc.). In general, all scientific measurements that you make should be in metric units. However, since you may occasionally obtain measurements in English units from another source,

a conversion table is provided at the end of this section for reference (Table 1.1). Since you have probably had some prior training with the metric system, this section will be brief and should serve as a review of the basic concepts of the metric system.

In the metric system the basic unit of length is the meter (m); the basic unit of volume is the liter (L); the basic unit of mass is the gram (g); and the basic unit of temperature is the degree Celsius (°C). The metric system is conveniently based on units of ten, which simplifies conversions from one metric unit to another. Simply moving a decimal point to the right or left is usually all that is needed to convert from one metric unit to another. Units of area are obtained by squaring the respective metric unit of length (e.g., 25 cm^2). Units of volume are obtained by either cubing the respective metric unit of length (e.g., 1 cm^3) or by measuring the displacement of the item in a fluid volume (e.g., 1 mL). The metric system was designed around the basic physical properties of water, one of the most abundant compounds on our planet. One gram (1 g) of water at 4°C occupies one cubic millimeter (1 mm^3) or one milliliter (1 mL) of volume. Fluid measurements in cubic centimeters (cm^3) are commonly abbreviated with the designation "cc". At sea level and under standard atmospheric pressure, water boils at 100°C and freezes at 0°C.

1. Obtain a meter stick or metric ruler from your instructor and measure the following items:

   Length:

   Pencil = _____ cm

   Staple = _____ mm

   Your height = _____ m

   Area:

   Credit card = _____ cm²

   Table surface = _____ m²

2. Obtain graduated cylinders of several sizes from your instructor and measure the volume of fluid that the following items can hold:

   Coffee cup = _____ mL

   Gallon jug = _____ L

NOTE: When reading the volume of fluid in a graduated cylinder, position your eyes level with the water line in the cylinder and record your measurement by aligning with the bottom of the meniscus (the curved surface of the fluid caused by surface tension between the fluid and the walls of the cylinder).

---

3. If an electronic balance or triple beam balance is available, record the mass of the following objects:

   Paper clip = _____ mg

   Quarter = _____ g

   Coffee cup (empty) = _____ kg

   Pencil = _____ g

4. Using Table 1.1 as a guide, convert the following measurements:

   187 cm = _____ m

   763 mm = _____ inches

   42 yd. = _____ m

   6.2 mi. = _____ km

   37.9 fl oz. = _____ mL

   4.7 L = _____ gallons

   4,845 g = _____ kg

   0.32 kg = _____ g

   32.3 lb. = _____ kg

   37°C = _____ °F

   –15°F = _____ °C

   100°F = _____ °C

TABLE 1.1 ▪ Conversion Table of Metric Units and Their English Equivalents

Units of Length	Units of Volume	Units of Mass	Units of Temperature
1 m = 39.4 in. = 1.1 yd. = 100 cm = 1000 mm = 0.001 km	1 liter = 1000 mL = 1000 cm³ = 2.1 pints = 0.26 gallons = 35 fl. oz.	1 g = mass of 1 cm³ of water at 4°C 1 g = 0.035 oz. = 0.001 kg	$°C = 5/9(°F - 32)$ $°F = (9/5)°C + 32$ Ex: 37°C = 98.6°F (body temp) 0°C = 32°F (H_2O f.p.) 100°C = 212°F (H_2O b.p.)
1 cm = 0.394 in. = 10 mm	1 mL = 0.035 fl. oz. = 1 cm³ (1 cc)	1 kg = 1000 g = 2.2 lb.	
1 mm = 0.0394 in.	1 gal = 3.85 liters	1 oz. = 28.35 g	
1 nm = 10^{-9} m = 10^{-6} mm	1 fl. oz. = 28.6 mL 1 qt. = 0.943 L	1 lb. = 0.45 kg	
1 yd. = 0.91 m			
1 ft. = 30.5 cm			
1 mi. = 1.61 km			
1 in. = 2.54 cm			

EXERCISE 1–2

Pipetting

A common task that you will perform in this laboratory is pipetting. A working knowledge of the different types of pipettes and pipettors is therefore essential to your success in the laboratory. The first rule of pipetting is: *Never pipette anything by mouth! Always* use a pipette bulb or pipettor on serological pipettes. There are three basic kinds of pipettes that you will encounter in the laboratory: disposable pipettes, serological pipettes, and micropipettors.

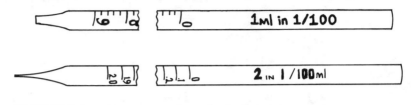

FIGURE 1.1 ▪ Two types of serological pipettes.

FIGURE 1.2 ▪ Pipette calibration. The first number indicates the pipette's total volume (in mL) and the second fractional number indicates the pipette's smallest calibrated increments. This is a 5.0 mL pipette divided into 0.1 mL increments.

DISPOSABLE PIPETTES

Disposable pipettes are generally made of polypropylene or some other flexible plastic and often contain calibration increments. They are, however, the least accurate of the three types of pipettes. Despite this shortcoming, they are the most commonly used pipette since they are quite handy for dispensing small, imprecise volumes quickly. The bulb and pipette are a one-piece unit which permit ease of use and allow for economical production, making them very inexpensive. Even though there are often calibrations, these marks should not be relied upon if very precise volumes must be delivered. Use disposable pipettes when speed is important and imprecision is tolerable (e.g., when dispensing an amoeba or planaria to a slide for wet mounting).

SEROLOGICAL PIPETTES

Serological pipettes are made of either glass or plastic and come in two types: those that are calibrated to deliver their entire volume by completely draining the pipette and blowing out the remaining drop in the tip (Fig. 1.1, upper diagram), and those in which the tip is not graduated (Fig. 1.1, lower diagram). This second type is called a Mohr pipette and is generally less commonly used. When you obtain a pipette, read the calibration numbers located on the neck (Fig. 1.2). The first number indicates the pipette's total volume (in mL) and the second number (usually a fraction) indicates the pipette's smallest calibrated increments. The pipette in Figure 1.2 is a 5.0 mL pipette divided into 0.1 mL increments.

To obtain a precise volume of fluid using the pipette, read the base of the **meniscus** (the curve created by surface tension from the sides of the pipette). Notice that the volume in the far left pipette in Figure 1.3 is slightly larger than 3.0 mL (3.1 mL), while the volume in the pipette at the far right is slightly smaller (2.9 mL). The pipette in the center of the diagram depicts the proper way to measure 3.0 mL using this pipette.

Serological pipettes are used in conjunction with some type of mechanical pipettor. Figure 1.4 depicts several mechanical pipettors which attach to serological pipettes and a micropipettor that is discussed in the next section. Your instructor will show you how to use the style of pipettor available in your laboratory.

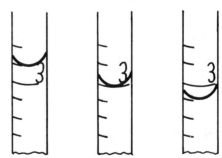

FIGURE 1.3 ▪ Determining the volume of fluid in a pipette by viewing the meniscus. The base of the meniscus should be level with the increment mark you are viewing on the pipette—as depicted in the middle pipette.

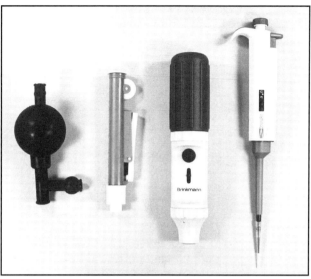

FIGURE 1.4 ▪ Common types of mechanical pipettors. From left to right: a simple rubber pipette bulb, a plastic pump with wheel and lever mechanisms, a pipettor with a rubber bulb and trigger mechanism, and a micropipettor.

MICROPIPETTORS

Micropipettors are used to precisely deliver very small volumes of fluid typically ranging from 1 μL to 1 mL. They use disposable tips which lack any calibration increments. Instead, micropipettors have a number stamped on them or a numerical dial that indicates volume. Typically this number is reported in microliters (μL), so a P-1000 would deliver 1000 μL (or 1 mL) of fluid. Some micropipettors deliver a fixed volume, while others are adjustable. As a group they are extremely delicate instruments which require precise calibration and should be handled carefully. Unless you are working with very small fluid volumes, you probably will not use micropipettors. If micropipettors are needed for an experimental protocol, be sure to ask your instructor for assistance before using one for the first time.

Check Your Progress

1. For each of the tasks below, select the pipette best suited for the job.
   a. Dispensing 4 μL of DNA extract into the loading wells of a gel.   _____
   b. Transferring 8 mL of bacterial culture from stock test tubes.   _____
   c. Placing 2 drops of stain on a tissue sample.   _____
   d. Decanting supernatant from a test tube of centrifuged cells.   _____

EXERCISE 1–3

Microscopy

Materials needed:
- compound microscope
- slide of letter "e"
- slide of crossed threads
- slide of ruler section *or* clear plastic ruler

The microscope is a tool that you will use repeatedly in the laboratory to discover minute structures and details that cannot be seen with the unaided eye.

Unfortunately, many students never refine their microscope skills to their maximum potential and are consistently missing a substantial portion of the material presented in laboratory exercises utilizing microscopes. It takes dozens of hours in front of the microscope to polish your technique to the point where everything you look at is in sharp contrast and clearly focused. Whether you have used a microscope before or not, carefully read through this section and perform the exercises listed. There's something for everyone to learn here. Your first goal is to familiarize yourself with the mechanical parts of the compound microscope.

Obtain a microscope from the cabinet. *Be sure to carry it with both hands* while transporting it to your laboratory bench. Microscopes are expensive, precision instruments that may be damaged easily by dropping or excessive jarring.

PARTS OF THE MICROSCOPE

Label the diagram of the microscope below with the terms for the major parts. Since there are numerous brands and models of microscopes available, some features on your scope may differ slightly from the one illustrated in Figure 1.5. Your instructor will point out any differences between your microscope and the one illustrated.

Photographic Atlas Reference Page 3

ocular lenses—lenses nearest the eye through which you look

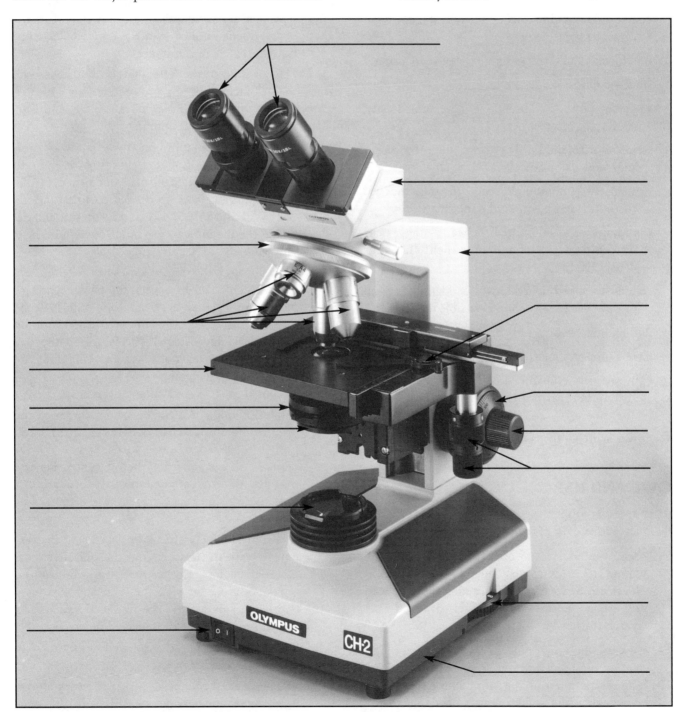

FIGURE 1.5 ▪ A compound binocular microscope.

objective lenses—lenses of different magnification that work in conjunction with ocular lenses to magnify the image

body—housing that keeps ocular and objective lenses in proper alignment

nosepiece—revolving housing that supports objective lenses

arm—supports microscope body, stage and adjustment knobs

coarse focus adjustment—moves stage up or down to focus image

fine focus adjustment—permits precise focusing

stage—supports slides

stage clips (may be absent)—hold slide in steady, stationary position

stage adjustment knobs—move stage to center slide under objective lens

condenser—lens mounted beneath stage that focuses the light beam on the specimen

iris diaphragm—mounted beneath stage near condenser; regulates amount of light illuminating specimen

condenser adjustment—moves condenser lens up or down to focus light (not visible on diagram)

illuminator—source of light

base—supports microscope unit

light intensity adjustment dial—rheostat (dimmer switch) that permits further adjustment of light intensity

power switch—turns microscope light on or off

CARE AND USE

After familiarizing yourself with the basic parts of the microscope, you're ready for some hands on use.

1. Clean the ocular and objective lenses by gently wiping them with dry lens paper. *Caution:* **Never use Kimwipes or other materials to clean microscope lenses! They are abrasive and may scratch the lens.**

2. Unwrap the cord, plug in the microscope and turn on the power switch. If you do not see any light coming from the illuminator, check to ensure that the light intensity slide near the power switch (if present) is not set to the lowest

setting. If that fails, alert your instructor—you may have a burned out bulb.

3. Rotate the nosepiece until the low-power objective "clicks" firmly into place directly over the stage. On most microscopes, the lowest power lens is either a 4X or 10X lens. The number will be stamped on the side of the lens.

4. Turn the coarse focus adjustment knob to completely raise the stage until there are only a few millimeters between the stage and the objective lens. On some microscopes, adjustment of the coarse focus knob moves the entire microscope body rather than the stage.

5. Place a slide of the letter "e" on the stage with the "e" facing upright and centered with the "e" directly over the hole in the stage and surrounded by a circle of light.

6. Adjust the iris diaphragm to the midway position allowing a moderate amount of light to penetrate the slide. You may open or close the iris diaphragm later as needed to fine tune the clarity of the image. As a rule of thumb, the higher the magnification you use, the more light you will need to see an image clearly.

7. Look through the ocular lenses at the letter "e". Adjust the distance between the ocular lenses to match the distance between your pupils. Resist the temptation to close one eye when looking through the microscope! This is not an effective alternative to using both eyes and will eliminate the advantages of binocular 3-D vision as well as contribute to eyestrain and even headaches.

8. While looking through the ocular lenses at the "e", rotate the coarse focus adjustment knob until the "e" comes into focus. If you do not see the "e" the slide may be off center. Be sure that the slide is aligned directly beneath the objective lens and held firmly by the stage clips (if present).

9. **Remember:** Always focus by *increasing* the distance between the stage and the objective lens. Never focus in the other direction—by decreasing the distance between the stage and the objective lens. This avoids the financial expense to you of broken slides or damaged objectives.

10. Use the fine focus adjustment knobs to finely focus the image in view.

11. Check to see if one of your ocular lenses has an independent focus adjustment. Usually one ocular will have a series of "tick marks" on the side

with a "0" and a "+" and "−" sign. When the "0" is aligned with the indicator mark on the side of the ocular lens, both ocular lenses have equivalent focal distances. If you wear contact lenses to correct your vision or have 20/20 vision, you will want both oculars to have the same focal plane. If you wear eyeglasses, it is recommended that you remove them when using the microscope, since its optical adjustments will allow you to correct for most visual problems.

12. Rotate the condenser adjustment knob to increase contrast and sharpen the image. The condenser is an often-overlooked, but extremely valuable component of the microscope which should not be neglected. You will find its use absolutely necessary to obtain sharp images with certain slides.

13. Adjust the position of the slide on the stage so that the image of the "e" is perfectly centered in your field of view. Only the central portion of the entire field of view will be visible as you increase magnification.

14. Carefully rotate the next objective lens (10X or 40X) until it clicks firmly into place *while viewing the slide from the side* to ensure that the lens does not contact the slide. ***Do not lower the stage to accommodate the next lens.*** If the slide is too thick to allow the next lens to swing into position, ***do not*** use that lens!

15. From this point on, use only the fine focus adjustment knob to focus the image.

16. Since microscope lenses are adjusted at the factory to be **parfocal**, the plane of focus and center of field of view should be nearly identical between different objective lenses. As a result, only minor adjustments should be necessary to bring the object into clear view and to center it in your field of view.

Check Your Progress

1. Describe the orientation of the letter "e" as it appears through the ocular lenses.

2. As you move the slide toward the right of the stage, to which direction does the image of the "e" move when viewed through the microscope?

3. As you move the slide away from you on the stage, to which direction does the image move?

USING HIGH-POWER LENSES

When using high-power objective lenses (40X and 100X), remember that the **working distance** between the objective lens and the slide is only a few millimeters or less. Therefore, the fine focus adjustment is sufficient to bring images into clear view. ***Never use the coarse focus adjustment with high power lenses.***

1. If your scope is equipped with a 40X lens, once the image is in focus under medium power, rotate the 40X objective into place, while viewing the slide from the side to check for adequate clearance. If the objective does not rotate into place without clearing the slide, do not force it. Ask your lab instructor to help you.

2. If you cannot locate the object through the ocular lenses and you find that the 40X objective is more than a centimeter away from the slide, you have passed the focal plane of the lens. Therefore, start from the beginning again with the lowest power objective and progressively work your way back to the 40X. Be sure that you do not lower the stage (or raise the objectives) once you

have focused the image and wish to change to the next objective.

3. Your scope may be equipped with a 100X lens. This is a special type of lens known as an oil immersion lens. The working distance is so small between this lens and the slide, and the magnification is so great, that it is necessary to place a drop of immersion oil between the coverslip of the slide and the 100X lens to obtain a clear image. At such high magnification the distortion of light rays caused by light passing through glass, then air, then into glass again is perceptible and causes a blurred image. Immersion oil has the same optical density as glass and thus stray light rays are not lost or distorted. *Do not attempt to use the 100X lens without assistance from your instructor.*

MAGNIFICATION

Together, the ocular and objective lenses constitute the magnifying system of your microscope. The initial magnification of the objective lens provides an image with good detail but is too small for easy examination. The ocular lens supplies secondary magnification of the initial image so that details are clear enough for normal viewing. The resulting image is a magnification of a magnified image and since the properties of magnification are multiplicative, you can easily calculate the total magnification of a specimen by multiplying together the independent magnification values of each lens. Remember that the magnification of objectives and oculars is stamped upon them.

Record values in Table 1.2 for the total magnification of each of your microscope lens pairs.

FIELD OF VIEW

The circular field that you see when looking through the microscope is described as the **field of view**. The diameter of this field of view changes with different magnifications. While still observing the slide of the letter "e" rotate the low, medium and high power objective lenses into place and compare the amount or proportion of the entire letter that is visible under each one.

TABLE 1.2 ▪ Compound Microscope Lens Magnifications

Objective Lens in Place	Objective Lens Magnification	Ocular Lens Magnification	Total Magnification
Low power	×	=	
Medium power	×	=	
High power	×	=	

Check Your Progress

1. Under which objective lens is the field of view largest?

2. Under which objective lens is the field of view smallest?

3. If you didn't know what you were already looking at, could you still determine it was an "e" using high power alone? How?

Knowing that a higher power lens has a smaller field of view is informative, but not precise enough to allow you to calculate the relative size of objects that you are viewing. To do this you must precisely know the diameter of each field of view for your microscope. An easy way to determine the size of the field of view of each objective lens is to place a clear plastic ruler (or prepared slide of a section of a ruler) on the stage, much like you would a slide, and view it through the ocular lenses.

1. Obtain a slide of a section of a ruler or small clear plastic ruler and place it on the stage.

2. Using low power, view the ruler through the ocular lenses and estimate the size of the field of view by measuring the diameter in millimeters. You should estimate this value to the nearest 0.1 mm.

3. Repeat this procedure with the medium power lens in place.

4. Since the ruler (or slide) is too thick to observe using the higher power lenses, *do not attempt to measure the field of view of the 40X or 100X lenses.* Your instructor can provide the values for the diameters of those lenses.

Check Your Progress

1. The diameter of the field of view of your low power lens is _____ mm.

2. The diameter of the field of view of your medium power lens is _____ mm.

3. Convert these values to micrometers (μm). Remember, there are 1,000 μm in 1 mm.

   a. low power lens = _____ μm

   b. medium power lens = _____ μm

4. As a rule, as magnification increases diameter of field of view _____.

DEPTH OF FIELD

The thickness of an image that is in focus at any point in time is referred to as the **depth of field** of a lens. Depth of field also varies with the magnification of the objective lens in place. You can establish the differences in depth of field of your objectives by viewing a slide containing overlapping objects.

1. Obtain a slide with a few strands of overlapping colored threads.

2. First view this slide on low power.

3. Concentrate on a section where the overlapping of the threads can be seen.

4. Are all three colored threads in focus using low power?

5. Now switch to medium power and try to determine which thread is on top of the other two. As you focus through the image, some threads will be in focus and others will be blurred.

6. Now switch to high power and repeat this procedure.

7. Can all three threads be in focus at the same time with the high power lens?

Check Your Progress

1. Which lens has the greatest depth of field?

2. As a rule, as magnification increases depth of field _____.

EXERCISE 1–4

Making a Wet Mount

> *Materials needed:*
> - glass slides
> - coverslips
> - pond water culture
> - compound microscope
> - disposable pipettes

A common way to view living organisms or tissues with the microscope is by making a **wet mount** of the specimen. This technique allows you to observe movements and properties of living specimens that are impossible to view with prepared slides. See Figure 1.6.

1. Place a drop of pond water culture in the center of a clean glass slide.

2. Carefully add a coverslip by placing one edge along the drop and gently lowering it onto the slide.

3. Press *gently* on the coverslip to remove any tiny air bubbles that may have been trapped in the process.

4. View this slide using low power.

5. Experiment with different intensities of light, and condenser and iris diaphragm settings, to maximize the clarity and contrast of the specimens in view.

6. Switch to higher magnifications and readjust condenser and iris diaphragm settings.

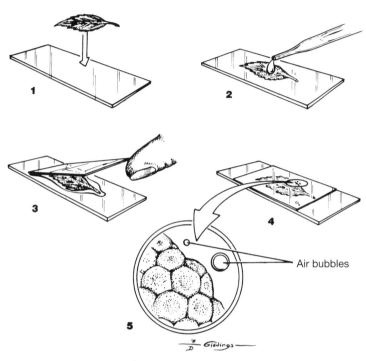

Air bubbles

FIGURE 1.6 ▪ Technique for preparing a wet mount of a biological specimen.

Check Your Progress

1. Which level of magnification requires the most illumination for the best clarity and contrast?

2. Examine your slide and sketch some of the organisms that you see in the space provided.
   NOTE: You may see a mixture of plant, animal and protozoan material on your slide.

3. Why is it imperative that you place a coverslip over the drop of fluid when making a wet mount?

EXERCISE 1–5

Basic Dissection Techniques

Since dissections will constitute a majority of the exercises in this laboratory manual, it is essential that you hone your dissection skills early on to be successful in this laboratory. A brief review of basic dissection techniques and suggestions will help build your proficiency, ensuring that you get the maximum benefit from your studies of the specimens detailed in this book.

1. Practice safe hygiene when dissecting. Wear appropriate protective clothing, gloves and eyewear, and *do not* place your hands near your mouth or eyes while handling preserved specimens. If fumes from your specimen irritate your eyes, ask your instructor about the availability of goggles.

2. Read all instructions *carefully* before making any incisions. Make sure you understand the direction and depth of the cuts to be made—many important structures may be damaged by careless or imprecise cutting.

3. Use scissors, a teasing needle and a blunt dissecting probe whenever possible. Despite their popularity, scalpels usually do more harm than good and should not be relied upon as your primary dissection tool. Remember the purpose of "blunt" dissection is to separate muscles, organs and glands from one another without cutting them.

4. Resist the temptation to stick your scalpel or teasing needles into the rubber or wax bottom of your dissecting pan. This unnecessarily dulls your instruments. Sharp tools are essential to performing clean, precise dissections.

5. When instructed to "expose" or "view" an organ, remove all of the membranous tissues that typically cover these organs (fat, fascia, etc.) and separate the "target" organ from neighboring structures. Your goal should be to expose the organ or structure as completely as possible.

6. A good strategy to use if working in pairs is to read aloud the directions from the book while your partner performs the dissection. These roles should be traded from section to section to give both of you a chance to participate.

7. Refer to illustrations and photographs frequently, but focus primarily on the specimen. Remember pictures are intended to help you in your dissections, but are not intended to substitute for the study of real specimens.

EXERCISE 1–6

Body Symmetry, Body Planes and Body Regions

> *Materials needed:*
> - preserved sponges, sea anemones, sea urchins, sand dollars, insects and vertebrates
> - orange
> - kitchen knife
> - permanent markers

Animals differ in their arrangement of body parts and these differences can be described best in relation to certain reference planes or axes. There are three major categories of body symmetry in animals: asymmetry, radial symmetry and bilateral symmetry (Fig. 1.7).

Asymmetry—lack of symmetry; irregular arrangement of body parts with no plane of symmetry to divide them into similar halves.

Radial Symmetry—body parts are arranged around a central axis; any plane passing through the central axis divides the body into two similar halves.

Bilateral Symmetry—body parts are divided into similar halves (mirror images) by a single plane of symmetry.

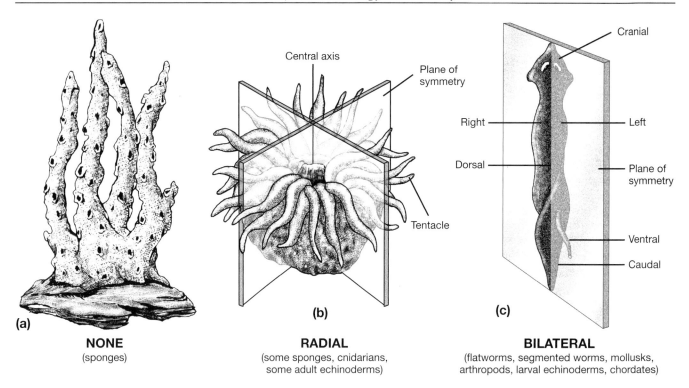

FIGURE 1.7 ▪ Patterns of body symmetry in animals: (a) asymmetry, (b) radial symmetry, and (c) bilateral symmetry.

Check Your Progress

1. Your instructor has set out a number of different organisms. Categorize each one according to its body symmetry.

   Asymmetrical:

   Radially symmetrical:

   Bilaterally symmetrical:

2. Which type of symmetry do you think is most prevalent among animals?

3. Radial symmetry is most common among sedentary organisms or organisms that drift passively with water currents. Speculate about the adaptive value of radial symmetry for these organisms.

Many anatomical references to body planes and regions differ between radially symmetrical and bilaterally symmetrical animals as well as between quadrupedal animals and bipedal animals (such as humans). For example, the ventral surface of a quadruped is equivalent to the anterior surface of a human and the oral surface of a jellyfish medusa. Because the majority of animals you will encounter in this manual are bilaterally symmetrical and quadrupedal, the following terms will be used to refer to the regions of the body and the orientation of the organs and structures you will identify.

A section perpendicular to the long axis of the body separating the animal into cranial and caudal portions is called a **transverse plane** (Fig. 1.8). The terms **cranial** and **caudal** refer to the head and tail regions, respectively. A longitudinal section separating the animal into right and left sides is called a **sagittal plane**. The sagittal plane running down the midline of the animal has a special name, the **median plane**. Structures that are closer to the median plane are referred to as **medial**. Structures farther from the median plane are referred to as **lateral**. **Dorsal** refers to the side of the body nearer the backbone, while **ventral** refers to the side of the body closer to the belly. A longitudinal section dividing the animal into dorsal and ventral parts is called a **frontal plane**. **Proximal** refers to a point of reference nearer the median plane or point of attachment on the body than another structure (e.g., when your arm is extended, your elbow is proximal to your hand). **Distal** refers to a point of reference farther from the body's median plane or point of attachment than another structure (e.g., when your arm is extended, your elbow is distal to your shoulder). **Rostral** refers to a point closer to the tip of the nose.

If you still feel a little confused about the terminology associated with animal body planes and regions, try the simple exercise below.

1. Obtain an orange and a permanent marker.

2. Sketch a representation of a face on one small region of the orange—no more than one half of the surface (Don't worry about your artistic abilities; this is purely for reference). You have just created an imaginary spherical organism that you will now systematically cut into pieces! See Figure 1.9.

3. Using a large kitchen knife, make a cut passing through the frontal plane of your imaginary "organism" separating it into dorsal and ventral halves. (In humans these would be referred to as posterior and anterior halves, respectively.)

4. Make another cut passing through the transverse plane of the ventral half (or anterior half)—the

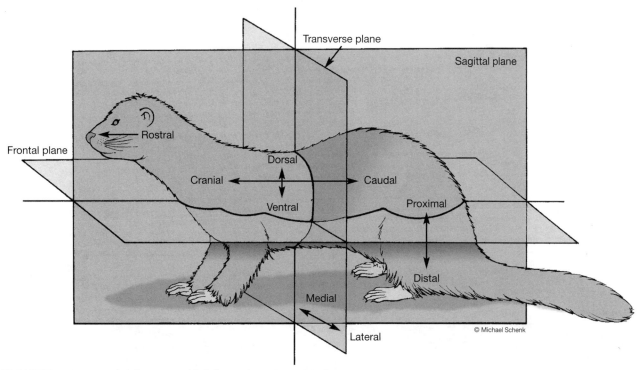

FIGURE 1.8 · Body planes and body regions in animals.

half with the face on it. This separates the ventral half of the orange into cranial and caudal quarters.

5. Finally, make a longitudinal cut through the sagittal plane of the remaining cranial quarter, separating the "face" into left and right halves through the median plane of the "organism."

FIGURE 1.9 ▪ "Hypothetical organism" depicting body planes and regions. First, make a vertical cut through your organism's frontal plane (1), then a horizontal cut through its transverse plane (2), and finally, another vertical cut through its sagittal plane (3).

Questions For Review

1. How many grams are there in 6.23 kg? _____

2. How much does 1 cm³ of water weigh? _____ g

3. Which distance is greater: 45.6 km or 24 miles? _____

4. List the three main types of pipettes.

5. If you have a microscope with a 5X ocular lens and a 40X objective lens, what is the total magnification of the image?

6. Describe what is meant by the term working distance.

7. Describe the change in each of the following as magnification decreases:

    a. field of view _____

    b. depth of field _____

    c. working distance _____

    d. light intensity requirements _____

8. Describe the procedure for making a wet mount.

9. List the 3 types of body symmetry and for each type list an animal that possesses that kind of symmetry.

10. Place the following terms in their correct locations in Figure 1.10 below.

cranial	medial	frontal plane
caudal	lateral	sagittal plane
dorsal	transverse plane	proximal
ventral	median plane	distal

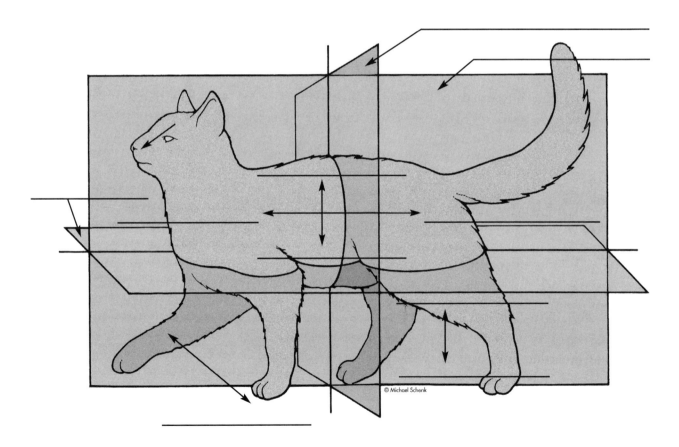

FIGURE 1.10 ▪ Unlabeled diagram of body planes and body regions.

Cells and Tissues

After completing the exercises in this chapter, you should:

1. Be familiar with the organelles of animal cells visible with the compound microscope.

2. Be able to discuss examples, functions and major histological features of each subclass of animal tissues.

3. Be able to give examples of where these tissues are found in the body.

4. Be able to define all boldface terms.

Before tackling the complexity of an entire living organism, it is useful to gain an understanding of the basic structural and functional units of animals—cells. The cell is the fundamental biological unit in living organisms and thus represents a tremendous milestone in the long evolutionary history of life on this planet. The present structure of the basic cell evolved some 3.5 billion years ago and since that time few changes have occurred in cells. Once a stable, functional and efficient design for a cell had arisen, most of the evolutionary changes occurred at the organismal level rather than the cellular or biochemical level.

An important concept in cell biology is that regardless of what kind of cell one is observing, certain functions and structures are indistinguishable.

Replication of cells is the same in an amoeba or a flatworm or a human. The mitochondrion from an earthworm is indistinguishable in its shape and function from the mitochondrion of a bird or a human. Similarities in cellular structure between such widely disparate groups of organisms illustrate how evolution has been very conservative at the biochemical and cellular levels. Despite the fact that cells have the same basic components and work under the same principles, cell exteriors have become markedly diverse in their appearance, reflecting their specialization to a myriad of different tasks and functions. To gain an appreciation for the variety of specializations that cells have undergone, you will examine several representative animal cells and tissues in the following exercises.

EXERCISE 2–1　　　　　　　　　　　　**Photographic Atlas Reference Page 3**

Animal Cells

> *Materials needed:*
> - slide of unfertilized sea star eggs
> - glass slides
> - coverslips
> - toothpicks
> - methylene blue stain
> - distilled H_2O (or isotonic saline solution)

1. Obtain a prepared slide of sea star eggs.

2. Use your low-power objective lens to locate a single, undivided spherical cell similar to the one illustrated in Figure 2.1.

3. After you have found such a cell, examine it using medium power then high power.

The three most easily identifiable parts of the cell are the central **nucleus**, the peripheral **cytoplasm** and the darker, surrounding **plasma membrane**. The nucleus is enclosed by the thin **nuclear membrane**, which regulates the passage of substances into and out of the nucleus. Within the nucleus are darkly-stained clumps of **chromatin** and a spherical structure called the **nucleolus** which manufactures ribosomes and exports them to the cytoplasm where they play a role in protein synthesis. Rarely will other cellular structures be visible with the light microscope. To view the many smaller subcellular organelles, an electron microscope is required.

After you feel comfortable identifying the basic parts of a cell on a prepared slide, you are ready to look at cells from your own body.

1. Use a toothpick to gently scrape the inside of your cheek.
2. Carefully swirl the scrapings in a small drop of distilled water (or isotonic saline solution) on a clean slide.
3. Add one small drop of methylene blue stain to the solution that now contains cheek epithelial cells.
4. Carefully lower a coverslip over the drop and gently press down on the coverslip to remove any air bubbles and flatten the cells.
5. Use the low power objective to locate a single flat cell or small group of cells.
6. View the cells using medium and finally high power.
7. Identify the same structures in these cells as you did in the sea star cell.
8. Sketch a picture of these cells in the space provided and label the parts that are visible.

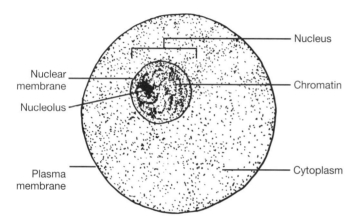

FIGURE 2.1 ▪ Generalized animal cell depicting structures visible with the light microscope.

Vertebrate Tissues

Materials needed:

- neuron slide
- simple squamous epithelium slide
- simple cuboidal epithelium slide
- simple columnar epithelium slide
- stratified squamous epithelium slide
- adipose tissue slide
- loose connective tissue slide
- hyaline cartilage slide
- elastic cartilage slide
- ground bone slide
- mammalian blood smear slide
- skeletal muscle slide
- smooth muscle slide
- cardiac muscle slide
- compound microscope

All animals are multicellular organisms and all but the simplest animals have their multitudes of cells arranged into discrete functional units called tissues. Sponges are a notable exception to this rule, since they lack true tissue organization. A **tissue** is defined as a group of similar cells that work together to perform a common function for an animal. Biologists classify animal tissues into four principal types, based on their structure and function:

Tissue Types

1. **Epithelial tissues**—cover external surfaces for protection or line the internal surfaces of body cavities and vessels.
2. **Connective tissues**—bind, support and protect body parts and systems.

3. **Muscle tissues**—permit movement of the animal through its environment and/or movement of substances through the animal.

4. **Nervous tissues**—initiate and transmit electrical nerve impulses to and from the body parts and store information in the form of biochemical compounds.

In many instances, these categories are further broken down in an attempt to compartmentalize the diversity of tissues that exist within each major group.

Photographic Atlas Reference Page 9

EPITHELIAL TISSUES

Epithelial tissues cover external surfaces for protection or line the internal surfaces of body cavities and vessels. They are typically arranged into tightly packed layers of cells with little or no intercellular space. They are further categorized based on the shapes of the cells and the number of layers of cells that constitute the tissue.

Simple Epithelium

Simple epithelial tissues consist of a single layer of cells and are classified based on their shapes. Squamous epithelium is comprised of flattened, irregularly shaped cells. They typically have a two-dimensional appearance in the microscope. When viewed from the side, they are often tricky to distinguish; a thin band of cytoplasm with a small bulge where the nucleus appears is usually all that is identifiable. **Simple squamous epithelium** would be represented by a single layer of flattened cells (Fig. 2.2a). The alveoli of lungs and the inner walls of arteries are comprised of simple squamous epithelium. Cuboidal epithelium and columnar epithelium contain cells that are thicker and fuller and have the three-dimensional appearances that their names suggest. Thus **simple cuboidal epithelium** would be represented by a single layer of cuboidal cells (Fig. 2.2b), while **simple columnar epithelium** would contain a single layer of columnar cells (Fig. 2.2c). Simple cuboidal epithelium is found in the tubules

of the mammalian kidney, while simple columnar cells are prevalent in the inner lining of the intestines in mammals. Epithelial tissues typically exist in simple layers when absorption or diffusion across the tissues is necessary.

1. Examine slides of simple squamous, simple cuboidal and simple columnar epithelium. Note the distinctive shapes of the cell types that make up each tissue.

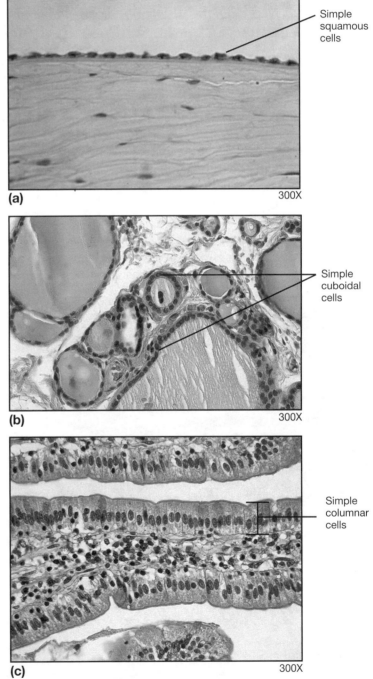

(a)   300X

Simple squamous cells

(b)   300X

Simple cuboidal cells

(c)   300X

Simple columnar cells

FIGURE 2.2 ▪ Epithelial tissues: (a) simple squamous, (b) simple cuboidal and (c) simple columnar.

2. Sketch representative examples of each of these three tissue types in the space provided below.

Stratified Epithelium

Stratified epithelial tissues derive their name from the layered appearance of the cells in these tissues (Fig. 2.3). In many cases, these tissues are comprised of more than one cell type (e.g., several layers of squamous cells followed by several layers of cuboidal cells). Epithelial tissues typically exist in stratified layers to serve as barriers against foreign substances and injury (e.g., the skin consists of an outer layer of stratified squamous epithelium to protect against abrasion and bacterial infection).

1. Examine a prepared slide of **stratified squamous epithelium**.

2. Sketch a representative sample of the tissue in the space below.

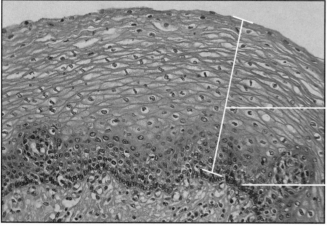

Stratified squamous layer

Basal layer

FIGURE 2.3 ▪ Stratified squamous epithelium.   200X

Check Your Progress

1. To what subgroup would you classify the cheek cells that you prepared earlier?

Photographic Atlas Reference Pages 10–11

CONNECTIVE TISSUES

Tissues that bind organs together, hold organs in place, support body structures and store food are grouped into the general category of connective tissues. Unlike epithelial tissues, connective tissues contain cells that are often widely separated by an extracellular matrix secreted by the living cells.

Cartilage

A common connective tissue in vertebrates is cartilage. Cartilage is comprised of a group of cells within a gelatinous **matrix** that provides firm, but flexible support. Embedded within this matrix are hollow chambers, called **lacunae**, which contain the **chondrocytes**, or cartilage-producing cells. **Hyaline cartilage** is found between bones, where it cushions the surfaces of joints. Its intercellular matrix is composed primarily of chondrin with thin collagen fibers to provide support and suppleness (Fig. 2.4a). **Elastic cartilage** contains fine collagen fibers and many elastic fibers which provide elasticity to this cartilage (Fig. 2.4b). This type of cartilage is much more flexible than hyaline cartilage and can be found in the ear, nose and voice box of humans.

1. Examine slides of hyaline and elastic cartilage. Notice the differences in appearance between the two types of cartilage.

2. Sketch representative examples of each type of cartilage in the space provided below.

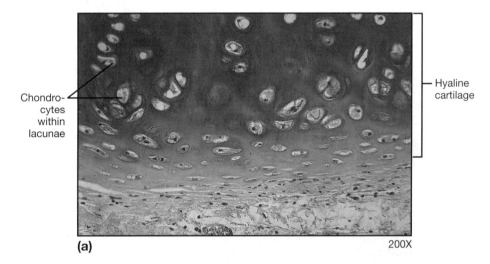

Chondrocytes within lacunae

Hyaline cartilage

(a) 200X

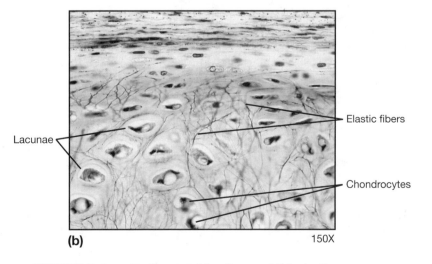

Lacunae

Elastic fibers

Chondrocytes

(b) 150X

FIGURE 2.4 ▪ Cartilage: (a) hyaline and (b) elastic.

Bone

Bone is another type of structural connective tissue. As bone grows, the **osteocytes** (bone producing cells) secrete a hard, calcified matrix which forms thin, concentric layers called **lamellae**, giving bone its characteristic appearance. These lamellae form layered rings around tiny, narrow pathways called **Haversian canals** (Fig. 2.5). As is the case in cartilage, the osteocytes are housed individually in **lacunae**. In living bone, the Haversian canals, appearing merely as black spots on a microscope slide, align themselves parallel to the long axis of the bone and contain blood vessels and nerve fibers to provide nourishment and information to the osteocytes. Nutrients are transported to each osteocyte through tiny, fingerlike projections in the lamellae called **canaliculi**, which form a miniscule canal system linking neighboring osteocytes for communication and nutrient transfer. In addition to providing structural support for the body, bone is responsible for storing calcium which can be withdrawn by the body as blood calcium levels drop, and for producing red blood cells in the bone marrow.

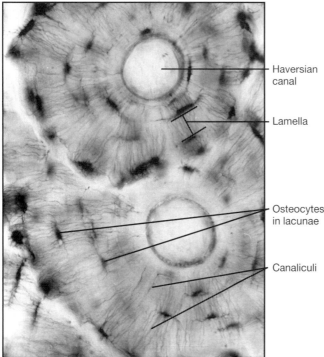

FIGURE 2.5 ▪ Bone tissue.   55X

1. Examine a slide of ground bone.
2. Sketch and label a small section of bone depicting a Haversian canal and several lamellae. Also label several lacunae and canaliculi.

Adipose Tissue

Adipose tissue is a type of connective tissue that stores or sequesters food for the body in the form of fat droplets. Each cell in adipose tissue is comprised of a large oil-filled vacuole, giving adipose cells the appearance of being empty spaces. These cells often become so full of fat that the small nucleus present is often pressed tightly against the outer margin of the cell membrane. As an adult, you have a finite number of fat cells in your body. Depending upon your diet and exercise regimen, these cells either enlarge as they store more fat or shrink as their energy stores are metabolized. Try as you might, you can never get rid of your fat cells through diet and exercise—you can only empty them of their contents and compress them into thin layers.

Loose Connective Tissue

Another connective tissue found in all vertebrates is **loose connective tissue**. Its name derives from its appearance—loosely scattered cells surrounded by a clear, gelatinous matrix (Fig. 2.6). Similar to the matrix found in cartilage, the loose connective tissue matrix contains thin elastic fibers and thicker, nonelastic fibers composed of collagen. This tissue is primarily responsible for holding other organ tissues together and in place in the body.

1. Examine slides of adipose and loose connective tissue.

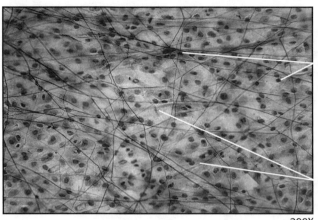

Elastic fibers (dark)

Collagen fibers (light)

200X

FIGURE 2.6 ▪ Loose connective tissue depicting elastic and collagen fibers.

2. Sketch a representative portion of each tissue below.

by cells. Red blood cells, or **erythrocytes**, appear as tiny, light pink, anucleate, concave discs, while the white blood cells, **leukocytes**, are larger and contain distinct, purplish nuclei (Fig. 2.7). Extremely small **platelets** should also be visible scattered among the other blood elements. Blood shares much in common with adipose tissue in that it too sequesters nutrients for the body within the plasma. Erythrocytes are the vehicles by which oxygen and carbon dioxide are relayed to and from the tissues, leukocytes are involved in the immune response, and platelets are the primary clotting agents of the blood.

1. Examine a prepared slide of blood.

Photo courtesy of Scott C. Miller

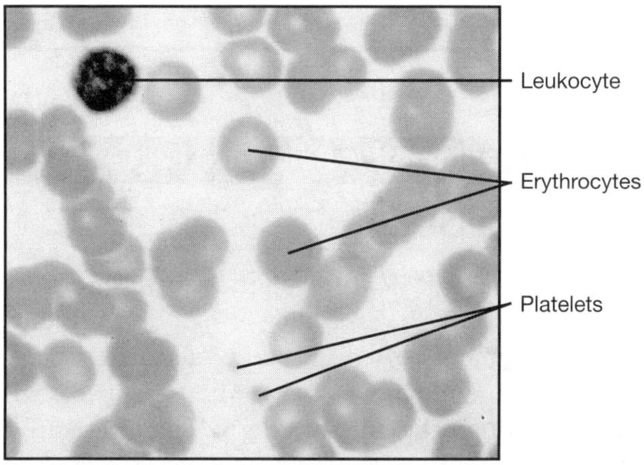

Leukocyte

Erythrocytes

Platelets

FIGURE 2.7 ▪ Fluid connective tissue: blood.

2. Sketch a small region of the slide of blood. Label the following: erythrocytes, leukocytes, platelets.

Blood

Blood, despite its fluid nature, is a type of connective tissue. Its cells and the fluid matrix, **plasma**, in which they are suspended coarse through blood vessels transporting oxygen, carbon dioxide, nutrients, electrolytes, hormones, metabolic wastes and practically any other substance used or produced

Check Your Progress

1. Which type of connective tissue provides the most rigid support?

2. Are nuclei visible in red blood cells?

3. Which type of blood cell is the most numerous? the least numerous?

4. What feature/s do all connective tissues share in common?

Photographic Atlas Reference Pages 11–12

MUSCLE TISSUES

The characteristic feature of muscle tissue is its ability to contract and thus create movement. These movements often propel an animal through its environment, but just as often propel substances through the animal's body. The interaction of **actin** and **myosin** filaments, which occur in abundance and in uniform orientation in muscle cells, is responsible for the contractility of muscle tissues. Three types of muscle tissue occur in vertebrates: smooth muscle, skeletal muscle and cardiac muscle. Similar to the case with adipose tissue, your body contains a finite number of muscle cells. As you exercise different muscles, the cells composing those muscles grow in response to the stress. What is perceived as a muscle getting bigger is actually the enlargement of existing cells, rather than the addition of new muscle cells.

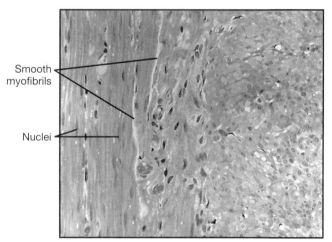

FIGURE 2.8 ▪ Smooth muscle tissue.   125X

Smooth Muscle

The simplest type of muscle tissue is smooth muscle. Lacking striations and generally confined to regions of the body under autonomic nervous control, **smooth muscle** fibers are long and spindle-shaped and contain single nuclei (Fig. 2.8). Smooth muscle tissue can be found in the bladder, the uterus, the stomach and in blood vessels. Contractions of smooth muscle are characteristically slow and rhythmic.

1. Examine a slide of smooth muscle.

2. From what region of the body was your sample of smooth muscle taken?

3. Sketch several smooth muscle cells.

2. Sketch several skeletal muscle cells. Label: nuclei, myofibrils, striations.

Skeletal Muscle

Skeletal muscle is composed of long, unbranched **myofibrils** which are actually composites of many individual muscle cells, giving these fibers their multinucleated appearance (Fig. 2.9). As skeletal muscles develop during embryonic development, many individual muscle cells fuse together and their nuclei get pushed toward the outer margins of the forming muscle fiber. This resulting "bundle" of cells enhances the strength and speed with which these fibers are able to contract. Skeletal muscle fibers make up the muscles attached to our skeletons and are under voluntary control. Skeletal muscle has a characteristic striated appearance, caused by the precise alignment of actin and myosin filaments along the **sarcomeres** of each muscle fiber.

1. Examine a slide of skeletal muscle.

Cardiac Muscle

Another type of striated muscle is found in the walls of the heart—**cardiac muscle**. Unlike skeletal muscle, nuclei are not located on the periphery of the cells and cardiac muscle is not under voluntary control; its steady, rhythmic contractions are controlled by a confluence of ganglia embedded in the muscle of the heart itself. Cardiac muscle is also composed of bands of muscle fibers which branch and reunite with one another to form a continuous network of muscle tissue (Fig. 2.10). The cells of these highly branched fibers are partially separated from one another by **intercalated discs**, a characteristic feature of cardiac muscle. These regions between individual cells appear as particularly

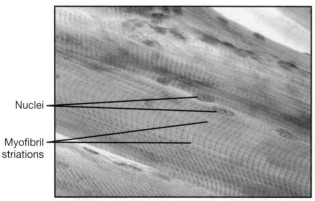

FIGURE 2.9 • Skeletal muscle tissue.   200X

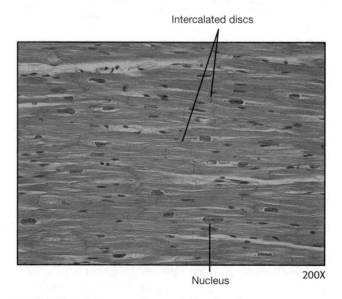

FIGURE 2.10 • Cardiac muscle tissue.

dark, bold striations in slides of cardiac muscle. Intercalated discs are gap junctions that allow communication between the cells of the muscle fibers and permit cardiac muscle to depolarize quickly and contract as a unit, much like skeletal muscle, though unlike the slow, wavelike contractions of other involuntary muscle.

1. Examine a slide of cardiac muscle.

2. Sketch a representative section of cardiac muscle. Label: nuclei, intercalated discs, striations.

Photographic Atlas Reference Pages 7–8

NERVOUS TISSUES

Nervous tissue consists of two major kinds of cells: neurons and supporting cells called glial cells. A **neuron** consists of: (1) a **cell body** containing the nucleus and other organelles, (2) a long **axon** which transmits electrical impulses away from the cell body and (3) short extensions called **dendrites** which typically receive electrical impulses from neighboring neurons or sensory receptors and transmit them to the cell body (Fig. 2.11). **Glial cells** assist in propagating nerve impulses and provide a nutritive role for neurons. Neuron cell bodies are located only in the brain and spinal cord; thus many axons must be of considerable length (some up to 1 meter!) to reach from the

spinal cord to the extremities of the body. In many areas, scores of axons are bundled together in cable-like nerve fibers to traverse the great distances required to reach the far corners of the body. The axons of these nerve cells have sheaths coated with a proteinaceous substance called **myelin**, giving them special electrical properties

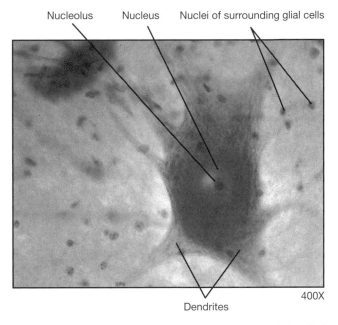

FIGURE 2.11 ▪ Nervous tissues: neuron with glial cells.

which greatly enhance the transmission speed of nerve impulses through these fibers.

1. Examine a slide of a neuron smear. Locate a single neuron in which the cell body, axon and dendrites are visible.

2. Sketch this neuron and its supporting glial cells below. Label the cell body, axon, dendrites and glial cells.

After viewing the slides of vertebrate tissues, fill in Table 2.1.

TABLE 2.1 ▪ Comparison of Vertebrate Tissues

Tissue Studied	Location in Body	Function

Questions For Review

1. Give an example of an epithelial tissue designed to "keep things out" of the body.

2. Give an example of an epithelial tissue designed to serve as a surface for diffusion of substances.

3. A common myth is that if you stop exercising your muscle will turn into fat. Explain why this is not true.

4. When your body metabolizes fat for energy it gets 9 Cal./g, as opposed to only 4 Cal./g from protein (muscle). Knowing this, explain why your body uses adipose tissue as the "default" energy store rather than muscle tissue.

5. Unlike mammalian red blood cells, the red blood cells of amphibians contain large nuclei. Speculate about the consequences of this on the relative oxygen-carrying capacity of amphibian red blood cells and the effects this could have on metabolic rates.

6. Describe two similarities and two differences between bone and cartilage.

Reproduction and Development

After completing the exercises in this chapter, you should be able to:

1. Recognize the stages of the cell cycle and understand the process of mitosis.
2. Describe major differences between mitosis and meiosis.
3. Understand the processes of spermatogenesis and oogenesis in vertebrates.
4. Understand the general sequence of embryonic development in deuterostomes.
5. Compare and contrast development in sea stars, sea urchins and frog embryos.
6. Define all boldface terms.

INSTRUCTOR'S NOTE: If you plan to do Exercise 3–4: *In Vitro* Fertilization with Sea Urchins, the exercise should be started at the beginning of class and monitored periodically throughout the remainder of lab to see each stage of cell division.

In the previous chapter you saw that cells work together as tissues to perform specific functions for animals. The process by which animals develop from a single fertilized egg cell into these complex, multicellular beings with dozens of different tissues specialized for vastly different roles is the topic of this chapter. You will first look at the mechanisms of cell division and **gametogenesis** (the production of egg and sperm cells) and then examine the process of embryonic development in two organisms with slightly different developmental pathways.

EXERCISE 3–1

Photographic Atlas Reference Pages 14–15

Mitosis

Materials needed:
- slide of whitefish blastula cells
- compound microscope

Throughout most of your life you grow (by adding more cells to your body), heal wounds and regenerate worn out or damaged tissues. All of these feats are accomplished by a process of cell division known as **mitosis**. Most cells in your body have the ability to replicate mitotically. Some cells, such as red blood cells, muscle cells and brain cells, sacrifice the ability to replicate when they become highly specialized for their specific tasks. For a cell to successfully replicate, it must leave an exact duplicate copy of its genetic material and an adequate supply of organelles in the "daughter" cell to allow that cell to function and read the genetic code it has been given.

The events that encompass the entire lifecycle of a cell from one division to the next are called the **cell cycle**. Mitosis occurs during only a portion of this cycle (usually a very small portion). The remainder of the cell cycle, **interphase**, is devoted to growth of the cell, synthesis of genetic material and other organelles, and taking care of the cellular duties which that particular cell is designed to do (e.g., absorb nutrients in the lining of the intestine, secrete saliva, or filter blood in the kidney tubules). The actual events of the cell cycle including mitosis are not discrete events, but occur in a continuous sequence. We separate mitosis and the cell cycle into discrete stages merely for the convenience of discussing and organizing this complex process.

You will examine a prepared slide of whitefish blastula cells arrested in mitotic development. This is an embryonic stage of fish development. When this embryo was preserved and sliced into thin sections, most of its cells were rapidly undergoing mitosis. Thus the slide you will examine contains an unusually high number of cells in mitotic division, which makes it ideal for studying mitosis.

1. Obtain a prepared slide of a whitefish blastula.

2. Use your low-power objective lens to locate a single section (or disc) of cells and examine this section carefully using high magnification (400X).

3. Locate several cells in interphase and each of the four stages of mitosis described below.

NOTE: You may see small purple dots in and around many of the cells. These dots are spheres of oil that serve as food packets for the developing embryonic cells. Do not confuse these spheres with chromosomes or other nuclear material.

4. If you do not see a nucleus or stained chromosomal material in the cell, do not attempt to categorize that cell. Often when the blastula is cut into extremely thin sections, the slice represented on the slide does not contain the nucleus or any nuclear material.

INTERPHASE

This stage, the longest of the cell cycle, is characterized by a distinct nuclear membrane enclosing lightly stained chromatin and a prominent nucleolus. At this point the DNA is loosely coiled into a thin spaghetti-like mass within the nucleus, often too dispersed to visualize (Fig. 3.1). DNA replication occurs during this phase. Centrioles also replicate, but remain invisible.

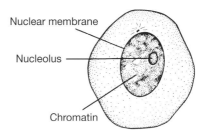

FIGURE 3.1 ▪ Interphase.

PROPHASE

The DNA coils tightly into chromosomes which appear as dark, rod-shaped structures. The nuclear membrane disintegrates and the nucleolus disappears. The two pairs of centrioles move to opposite poles of the cell and form a proteinaceous matrix of microtubules between them that will become the **spindle fibers** (Fig. 3.2). When the centriole pairs reach opposite ends of the cell, they anchor themselves to the cell membrane with an array of microtubule filaments known as **asters** radiating outward from the centrioles.

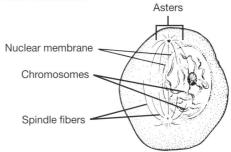

FIGURE 3.2 ▪ Prophase.

METAPHASE

The chromosome pairs align along the equatorial plane of the cell and each homologous pair of chromosomes attaches to a spindle fiber at its **kinetochore** (a site of attachment at the chromosome's **centromere**). Spindle fibers and asters surrounding the centriole pairs are typically visible at this stage of mitosis (Fig. 3.3).

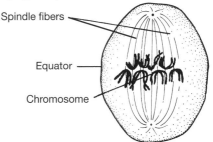

FIGURE 3.3 ▪ Metaphase.

ANAPHASE

The centromeres divide, separating single-stranded chromosomes from one another. These chromosomes are pulled toward opposite ends of the cell through contractions of the microtubules comprising the spindle fibers (Fig. 3.4).

FIGURE 3.4 ▪ Anaphase.

TELOPHASE

Single-stranded chromosomes continue their migration toward opposite poles and a distinct **cleavage furrow** develops along the equatorial plane of the cell. As telophase ends and cytokinesis (cytoplasmic division) begins, the microtubule networks begin to collapse and the spindle fibers disintegrate, the nucleolus reappears and a nuclear membrane forms around each of the two bodies of chromosomal material (Fig. 3.5). The chromosomes slowly disperse into their stringy, uncoiled state and the cell completes division of the cytoplasm.

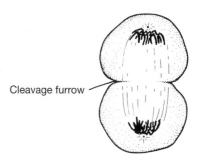

Cleavage furrow

FIGURE 3.5 ▪ Telophase.

Check Your Progress

1. If a cell has 20 chromosomes during early interphase, how many chromosomes would be present during prophase?

2. How many chromosomes would each of the two new cells have that resulted from the mitotic division of this cell?

3. The different stages of mitosis are not of equal duration; some stages are relatively long while others are quite brief. Since the frequency of cells in each stage is proportional to the relative duration of that stage, you can estimate the relative length of a stage by counting the number of cells arrested in that particular stage. Based on the number of cells you observed in each stage of mitosis (excluding interphase), which stage do you think takes the longest time to complete? (You may want to review the slide again and count a random sample of 50 cells to answer this question.)

4. Which stage takes the shortest time to complete?

Gametogenesis

> *Materials needed:*
> - testis slide
> - ovary slide
> - compound microscope

Not every cell in our bodies divides into two identical replicas of itself. If that happened in the reproductive cells of species that reproduced sexually, each generation of offspring would have twice the number of chromosomes in its cells as the previous generation! Biologically this is an intolerable situation for all but a few unique species. Some animals have evolved tolerances to tetraploid (4 times the normal number of chromosomes) or other chromosome combinations. Clearly then, the mechanism of cell division that produces egg and sperm cells is fundamentally different from the purely mitotic division responsible for growth and maintenance of our bodies. Meiosis produces **haploid gametes** (cells that contain half the number of chromosomes as the somatic cells) through the process of gametogenesis. Fertilization occurs later, restoring the genetic compliment of the zygote to the diploid condition.

Meiosis generally consists of two successive nuclear divisions known as first and second meiosis (or meiosis I and meiosis II). In **meiosis I**, the replicated, double-stranded chromosomes align along the equator of the cell and the homologous pairs of double-stranded chromosomes actually join together forming **tetrads**. During this period, **crossing over** occurs—a process in which small sections of chromosome are exchanged between neighboring homologous chromosomes, shuffling some of the genetic information between chromosome pairs. These homologous chromosomes separate from one another and move to opposite poles. The cell divides creating two haploid daughter cells which do not undergo replication of the genetic material. Then in **meiosis II**, the double-stranded chromosomes in each daughter cell divide again, this time separating into single-stranded chromosomes which move to opposite poles, producing four haploid daughter cells, each with half the number of single-stranded chromosomes as the original cell. Although the process of gametogenesis follows this same basic pathway in males and females, there are significant differences.

Therefore we will examine gametogenesis in the two sexes separately.

SPERMATOGENESIS

Spermatogenesis is the meiotic production of sperm cells, which in mammals occurs within the tiny, coiled seminiferous tubules of the testes. The process begins when diploid **spermatogonia** (singular = spermatogonium) divide mitotically to produce diploid **primary spermatocytes** (Fig. 3.6). These primary spermatocytes undergo meiosis I to become haploid **secondary spermatocytes**. The second meiotic division transforms each secondary spermatocyte into two haploid **spermatids**. At this point, the spermatids differentiate into **spermatozoa**, mature sperm cells, and are stored in the tubules of each epididymis that cups around a testis. From each original spermatogonium, four haploid spermatozoa are produced.

1. Observe a slide of a testis using high magnification and locate several sections through the seminiferous tubules (Fig. 3.7).

2. Notice that primary spermatocytes and spermatogonia are located near the periphery of the tubules, while maturing sperm are located near the middle of the tubules. As sperm mature, they migrate inward toward the center of the tubules for transport to the epididymis.

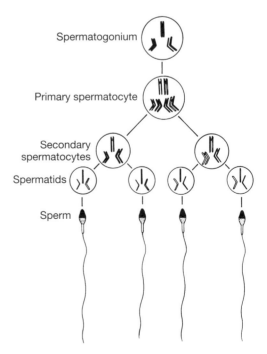

FIGURE 3.6 ▪ Spermatogenesis.

Spermatids differ from one another genetically because random segregation and assortment of the chromosomes during metaphase I of meiosis combined with the process of crossing over shuffle the parent genomes. Any given spermatid contains some maternal and some paternal chromosomal material —the particular combination is purely a matter of chance.

In human males, spermatogonia retain their ability to divide mitotically throughout life. Thus sperm production occurs in a continuous cycle from the time of sexual maturation until death, taking between 65–70 days for mature sperm to develop from spermatogonia. Somewhere on the order of 12 million new sperm are produced every day in the testes. In the biological sense, sperm are cheap—they require

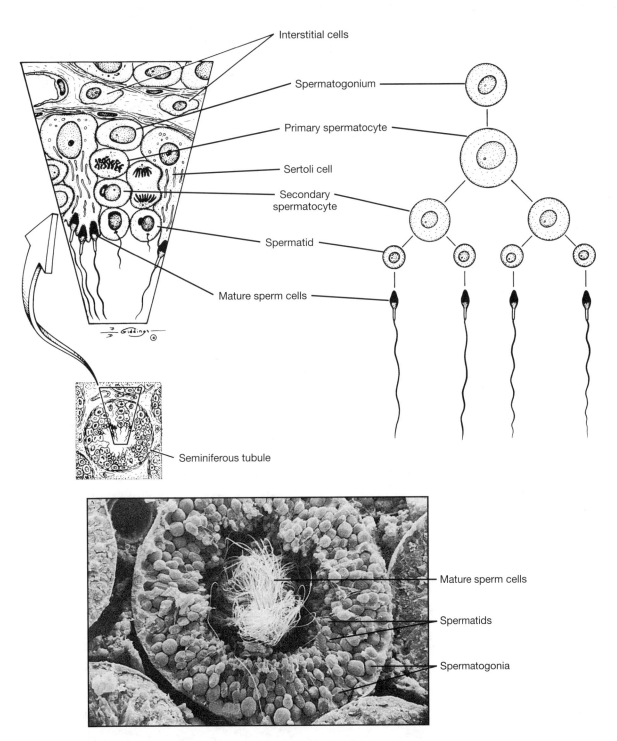

FIGURE 3.7 ▪ Section through mammalian seminiferous tubule showing spermatogenesis.

very little energy investment to produce and are in a virtually infinite supply throughout life. This represents quite a marked difference from the human egg as we will see next.

OOGENESIS

The meiotic production of eggs (or ova) is known as **oogenesis**. Within the ovaries, diploid **oogonia** (singular = oogonium) divide mitotically to produce diploid **primary oocytes** (Fig. 3.8). These primary oocytes immediately enter prophase I of meiosis, at which point their development arrests. Primary oocytes may remain arrested in prophase I of meiosis for days, weeks or even years! Only after the primary oocyte receives appropriate hormonal signals from the body will it progress through the latter stages of meiosis I. Contrary to what happens in spermatogenesis, four mature eggs are not produced from one primary oocyte. The first nuclear division during meiosis I occurs near the margin of the cell, rather than at the cell's equator, and grossly unequal portions of cytoplasm are allocated to each of the

resulting cells. The daughter cell receiving the bulk of this cytoplasm becomes a haploid **secondary oocyte**, while the one that receives virtually no cytoplasm forms the first **polar body**. Quite often, this is the stage at which fertilization by the sperm occurs. In many organisms, the second phase of meiosis will not be initiated unless fertilization does occur. The second meiotic division, which occurs in the secondary oocyte, also involves an asymmetric allocation of cytoplasmic resources and results in a large, haploid **ootid** (which will later become the ovum) and a second, nonfunctional polar body. Occasionally the first polar body proceeds through meiosis II and divides into two more nonfunctional polar bodies. Eventually the polar bodies disintegrate leaving only a large ovum well-stocked with nutrients and reserves to make it through the first few divisions after fertilization.

1. Observe a slide of an ovary using low or medium power magnification.

2. Locate ova in different stages of development.

3. Notice that as ova mature and the follicle surrounding them enlarges, they migrate outward, toward the periphery of the ovary (contrary to the pathway of maturing sperm cells). At the point of ovulation, the ovum will erupt from the ovary and be swept into an oviduct by currents created by cilia lining each oviduct (Fig. 3.9).

In human females, the proliferation of oogonia is limited to a short period of time before birth. Thus when a woman is born, she contains all of the primary follicles she will ever have available to her. This finite number of egg precursors numbers around 400,000 at birth, but only a few hundred of these will ever mature into ova and be released into the oviducts. Couple this with the fact that each egg is an extremely large cell packed full of cytoplasm containing nutrients, mitochondria and other organelles necessary for the early stages of embryonic development and you can see that unlike sperm, eggs are not cheap! They exist in a limited supply and require large energetic investments to produce and release.

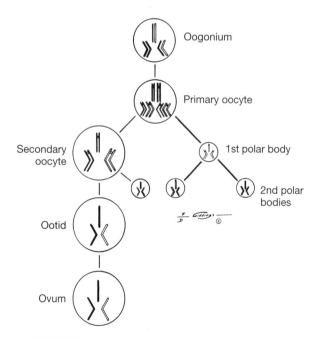

FIGURE 3.8 ▪ Oogenesis.

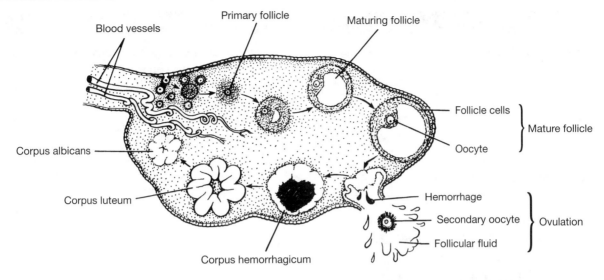

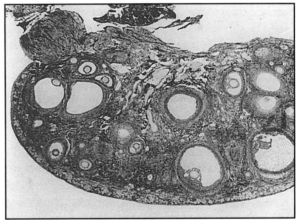

Photomicrograph of
mammalian ovary (100x).
Courtesy of Turtox, Inc.

FIGURE 3.9 ▪ Section through mammalian
ovary showing follicle development, ovulation
and formation of the corpus luteum.

Check Your Progress

1. How does meiosis create genetic diversity for natural selection to operate?

2. If a spermatogonium has 20 chromosomes, how many chromosomes will be present in a mature sperm cell from that spermatogonium?

3. How many sperm cells will one spermatogonium produce?

4. If an oogonium has 20 chromosomes, how many chromosomes will be present in an ovum from that oogonium?

5. How many ova will one oogonium produce?

6. Describe how the production of nonfunctional polar bodies is adaptive for the resulting ovum.

Animal Development

> *Materials needed:*
> - prepared slide of sea star developmental stages
> - prepared slide series of amphibian developmental stages
> - compound microscope

During the process of sexual reproduction, sperm and egg fuse to form a new cell. Typically the nuclear material from the haploid sperm is donated to and combines with the haploid nuclear material of the egg to restore the diploid condition in the fertilized egg or zygote. From this point, the single cell develops into a magnificently complex, multicellular organism capable of carrying out all the daily functions of life on its own. Three major processes govern the developmental sequence of animals: (1) **cell division** which creates new cells through mitosis, (2) **cell migration** which causes groups of cells to organize and move to create an animal's body shape and (3) **cell differentiation** which causes cells to develop different morphological features for specialization toward unique tasks. In the following exercise you will examine the development of two different animals, a sea star and a frog. They share many developmental traits in common, since they are both **deuterostomes** (the mouth develops from the second embryonic opening). However, they do possess variations in their developmental processes as well.

SEA STAR DEVELOPMENT

Sea stars of the genus *Asterias* are common marine invertebrates belonging to the phylum Echinodermata (see Chapter 12). These animals are often found along intertidal zones in the oceans and are the familiar creatures we often refer to as "starfish." Because of the many similarities in echinoderm and chordate developmental pathways, sea stars and sea urchins are common laboratory models for the study of animal development.

1. Obtain a prepared slide of *Asterias* containing a series of sea star developmental stages. As you scan the slide using low power, you will notice that your slide contains dozens of embryos in different stages of development. Identify each of the stages listed below using medium or high magnification.

2. **Unfertilized egg.** In this stage, the egg appears as a large, nearly spherical cell with a distinct nucleus and nucleolus (Fig. 3.10). Tiny granules are visible within the cytoplasm of the cell. Notice how these granules are uniformly distributed throughout the cytoplasm. These are particles of yolk which represent food for the developing embryo. Eggs with relatively little yolk that is evenly distributed throughout the cytoplasm are classified as **isolecithal**.

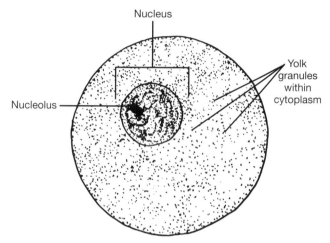

FIGURE 3.10 ▪ Unfertilized sea star egg.

3. **Fertilized egg.** After the sperm penetrates the egg, a thin **fertilization membrane** is secreted by the egg, ensphering it to prevent multiple fertilizations by the many nearby sperm. The nucleus is no longer visible as the nuclear material from sperm and egg fuse to restore the zygote to the diploid chromosome number (Fig. 3.11).

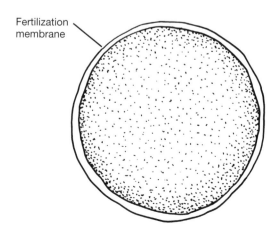

FIGURE 3.11 ▪ Sea star zygote (fertilized egg).

4. **Two-cell through eight-cell stages.** Shortly after fertilization, the zygote begins to divide through mitosis. Notice that the pattern of **cleavage** (cell division) completely separates the cells into roughly equal-sized, but distinct cells (Fig. 3.12). This pattern of development is known as **holoblastic cleavage** and is common among eggs with little or no yolk. Notice also that the fertilization membrane remains intact through these early cleavages. At this early stage of development it is still possible for nearby sperm to penetrate these cells in their relentless pursuit to fertilize the "egg" and consequently upset the genetic balance of the zygote. The persistence of the fertilization membrane prevents this occurrence. Finally, notice that as these cells divide they do not get bigger. The existing cytoplasm is partitioned equally among the new cells with very little growth.

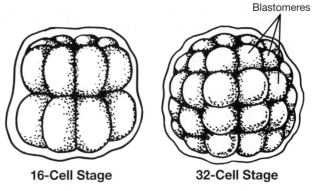

16-Cell Stage   **32-Cell Stage**

FIGURE 3.13 ▪ Morula of sea star.

6. **Blastula** (64+ cell stages). The blastula stage is characterized by a migration of cells toward the periphery of the embryonic sphere creating a hollow cavity deep within the spheroid of cells. At this stage, the blastula breaks free from the envelopment of the fertilization membrane and is able to swim about freely due to the ciliated blastomeres on the surface of the embryo. The hollow, fluid-filled cavity within the blastula is referred to as the **blastocoel** (Fig. 3.14).

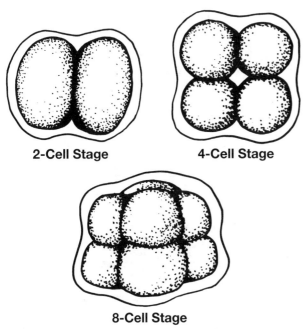

2-Cell Stage   **4-Cell Stage**

8-Cell Stage

FIGURE 3.12 ▪ Two-cell through eight-cell stages.

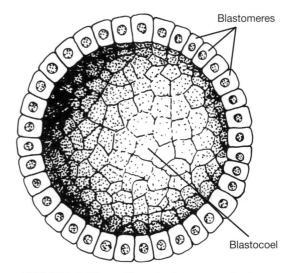

FIGURE 3.14 ▪ Blastula (cross-section).

5. **Morula** (16–32 cell stages). At this stage the embryo is represented by a solid mass of tiny cells (**blastomeres**) that continue to divide at a furious pace. Still the overall size of the embryo has not increased—rather the size of the individual blastomeres has decreased to accommodate the fixed amount of cytoplasm being divided among cells. The fertilization membrane is still usually visible at this stage (Fig. 3.13).

7. **Early gastrula.** Several hours after the formation of the blastula, a small depression begins to appear at one end of the embryo. This invagination of cells toward the center of the blastocoel marks the onset of gastrulation. As gastrulation proceeds, more and more cells stream inward, deepening the invagination. The opening to the outside of the embryo, marking the site of the inward migration of cells, is called the **blastopore** (Fig. 3.15). In sea stars and other deuterostomes this

opening will become the anus. The hollow tube that is created by the arrangement of invaginated cells is called the **gastrocoel** (or archenteron, meaning primitive gut).

8. **Late gastrula.** By now, the individual cells are so small as to be indistinguishable from each other. At most they appear as a flecking on the surface of the embryo. The archenteron has folded inward to the point of nearly connecting with the opposite end of the gastrula and the shape of the gastrula has become markedly elongated. No longer reminiscent of a spherical ball of cells, the gastrula continues to develop toward the larval stage. During gastrulation cell differentiation begins to form the primary tissue layers of the embryo. The outer layer begins to develop into **ectoderm**, while the inner layer differentiates into **endoderm** (Fig. 3.15). The third embryonic layer, **mesoderm**, will develop later between these two existing layers from cells that disassociate from the endodermal layer.

9. **Bipinnaria larva.** By this point **morphogenesis** (the development of body shape) and **organogenesis** (the differentiation of organ tissues) have begun. The larval sea star has a complete digestive tract including an anus and a functional mouth lined with cilia for sweeping in organic particles (Fig. 3.16). Soon after this stage, the young sea star will undergo another developmental transformation as it grows arms and assumes the familiar shape of an adult starfish.

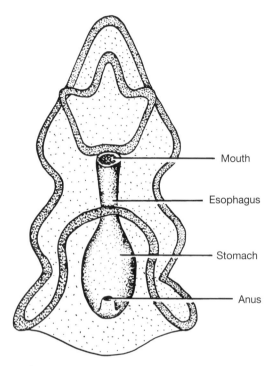

FIGURE 3.16 ▪ Bipinnaria larva of sea star.

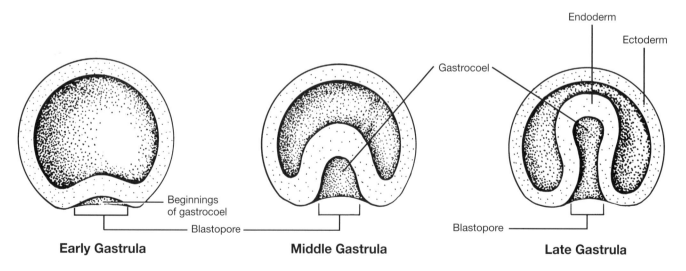

FIGURE 3.15 ▪ Gastrulation in the sea star.

FROG DEVELOPMENT

Frogs belong to the vertebrate class Amphibia within the phylum Chordata (see Chapter 15). Females of many species produce relatively large eggs (around 2 mm in diameter) and fertilization and embryonic development occur externally, making the frog another useful model for studying developmental pathways.

1. Examine prepared slides of the following stages of frog embryonic development. Models of these developmental stages or plastic-embedded specimens may also be available for study.

2. **Unfertilized egg.** Frog eggs are **mesolecithal**, meaning they contain a moderate amount of yolk. This yolk is segregated toward one pole of the egg known as the **vegetal pole** which appears lighter in coloration. The darker pole, or **animal pole**, represents the portion of the egg where the embryo will develop. Frog eggs are also enclosed in a gelatinous covering to protect them from bacterial infection, UV radiation, desiccation and to deter predation. The entire course of development takes place with the embryo encased in this coating of jelly. It is not until the embryo reaches the tadpole stage that it breaks free of its protective barrier and swims away.

3. **Fertilized egg.** When fertilization occurs, the **fertilization membrane** lifts away from the surface of the egg and the egg rotates so that the heavier, yolk-filled vegetal pole is downward. Later, a **gray crescent** appears along the margin of the animal-vegetal axis on the opposite side of the egg from the entry point of the sperm (Fig. 3.17). In many cases this feat of counter-alignment involves rotation of these pigmented zones, since the sperm may penetrate the egg at any place along its surface.

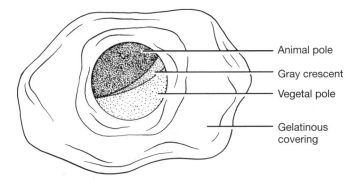

FIGURE 3.17 ▪ Frog zygote (fertilized egg).

4. **Early cleavage.** Cleavage in the frog embryo is **holoblastic**, similar to that in the sea star. The first two cleavage planes are perpendicular to one another producing four cells of roughly equal size (Fig. 3.18). The third cleavage occurs parallel to, but slightly above the equator that separates the animal and vegetal poles. From this point forward, the division of cells will follow this unequal pattern with cells in the animal pole dividing at a faster rate than cells in the vegetal pole. Vegetal pole cells divide more slowly because they are laden with yolk.

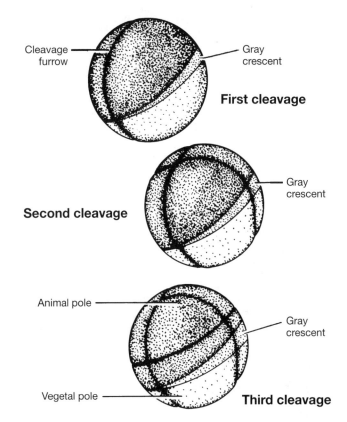

FIGURE 3.18 ▪ Early cleavage events in the embryonic development of a frog.

5. **Late cleavage.** By the later stages of cleavage, there is a marked difference in the appearance of the two poles. The animal pole is comprised of considerably smaller, more numerous cells while the vegetal pole contains relatively few, large cells containing yolk granules (Fig. 3.19).

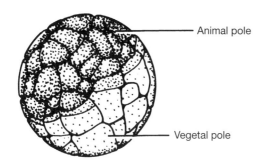

FIGURE 3.19 • Late cleavage in the embryonic development of a frog.

6. **Blastula.** With the formation of the **blastocoel**, the blastula stage is achieved. Notice that the blastocoel in the frog embryo is not centrally located as it is in the sea star. It is offset toward the animal pole of the embryo due to the presence of the large yolk cells of the vegetal pole (Fig. 3.20).

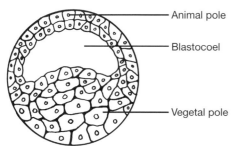

Longitudinal section, blastula

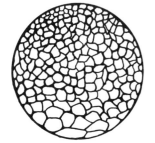

Intact late blastula

FIGURE 3.20 • Formation of the blastula during embryonic development in a frog.

7. **Gastrula.** The onset of gastrulation is characterized by a migration of surface cells inward toward the blastocoel. In the frog, this forms a crescent-shaped line along the surface of the blastula known as the **dorsal lip** which marks the opening of the **blastopore** (Fig. 3.21). As surface cells migrate inward, a slight depression forms on the surface of the blastula. This depression gradually enlarges to form the **archenteron** within the gastrula. Due to the large size of vegetal pole cells, blastomeres do not invaginate as they do in the sea star; rather they grow down over the larger yolk cells enveloping them as development progresses. By the late gastrula stage the vegetal pole cells have become almost completely enveloped by migrating surface cells from the animal pole and only a small, circular **yolk plug** remains visible at the opening to the blastopore. By this stage, the three primary **germ layers** have begun to differentiate. The outer surface of the gastrula is covered by **ectoderm**, the archenteron (gastrocoel) is lined with **endoderm** and the thin layer of cells sandwiched between the ectoderm and endoderm constitutes **mesoderm**.

8. **Neurula.** Several hours after gastrulation, the first visible elements of the nervous system begin to appear as a result of ectodermal cells along the mid-dorsal region of the embryo thickening to

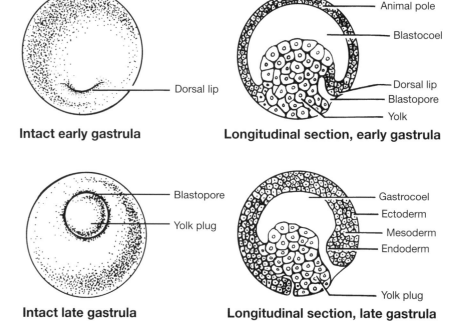

Intact early gastrula

Longitudinal section, early gastrula

Intact late gastrula

Longitudinal section, late gastrula

FIGURE 3.21 • Gastrulation in the frog embryo.

form two enlarged ridges on the surface, the **neural folds**, which border a depression, the **neural groove** (Fig. 3.22). This marks the onset of **neurulation**. Eventually the neural folds will meet and fuse together forming the enclosed **neural tube** which will develop into the brain and spinal cord. Below the developing neural tube, a long, cylindrical section of mesodermal cells is differentiating to form the **notochord** which will later develop into vertebrae.

9. **Larva.** Several days after fertilization the embryo begins to elongate and takes on the more familiar tadpole shape (morphogenesis). The caudal end develops a pronounced **tail** while **gills** and a **mouth** develop at the cranial end. What is left of the **yolk mass** remains as a slight bulge in the abdomen of the developing tadpole (Fig. 3.23). Once the tadpole emerges from its gelatinous enclosure, the yolk reserves will be depleted and it must find food for itself for the next several weeks (or months, depending on the species) until it has stored enough energy to metamorphose into a young frog.

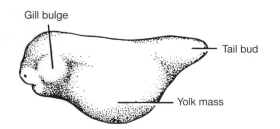

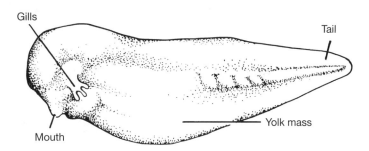

FIGURE 3.23 ▪ Early larval stages of frog embryonic development.

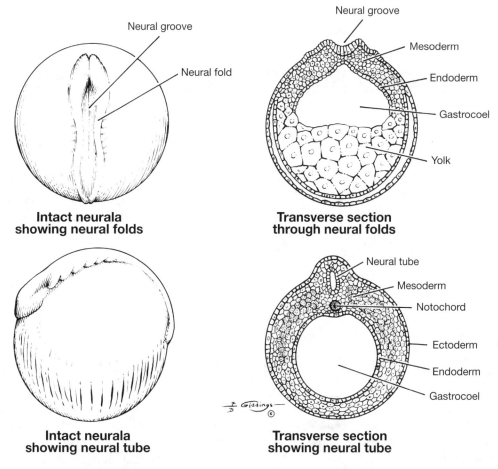

FIGURE 3.22 ▪ Neurulation in the frog embryo.

Check Your Progress

1. Which is larger, a sea star zygote or a frog zygote?

2. How does the pattern of cleavage differ in sea stars and frogs?

3. How does the formation of the blastocoel differ in sea stars and frogs?

4. How does the process of gastrulation differ in sea stars and frogs?

EXERCISE 3–4

In Vitro Fertilization with Sea Urchins

Materials needed:
- live sea urchins
- 35 mL petri dishes (sterile)
- 50 mL and 100 mL beakers (sterile)
- sterile sea water
- 4% potassium chloride solution (0.5 M)
- 5 mL syringes with 1/2 in. 25-gauge needles
- sterile pipettes
- toothpicks
- glass depression slides
- coverslips
- compound microscope

Much like sea stars, sea urchins have been extremely useful organisms for embryologists studying the process and mechanisms of development. Earlier, you viewed prepared slides of vertebrate eggs and sperm and developing sea stars. In this exercise you will actually perform *in vitro* fertilization (i.e., outside the body) using sea urchin eggs and sperm to create zygotes whose development you will track for the next several days. **Fertilization** is a complex set of steps involving: (1) contact between sperm and egg, (2) entry of the sperm into the egg, (3) fusion of the egg and sperm pronuclei which restores the diploid condition and results in the formation of a zygote and (4) activation of developmental mechanisms which initiate successive divisions of the new cell (by means of mitosis) to form the multicellular embryo. This process occurs naturally with great ease and miraculously with clock-like precision millions of times a day throughout the animal world. Unfortunately, to perform this exercise in the unnatural conditions of the laboratory requires painstaking attention to detail and protocol. To achieve successful fertilizations that can be monitored for several days, you will need to maintain an aseptic environment that mimics the natural conditions in which these animals would normally develop.

PRECAUTIONS FOR SUCCESSFUL DEVELOPMENT

1. **Cleanliness.** Use only dishes, pipettes, beakers, sea water, etc., that are sterile (either new or recently autoclaved). Do not cross-contaminate containers by using a pipette for more than one container. Label your containers and pipettes to avoid confusion!

2. **Avoid Overinsemination.** Very few sperm are necessary for successful fertilization. Use a _dry_ toothpick to introduce sperm. If the egg suspension becomes noticeably milky after adding sperm, you have added too many.

3. **Avoid Overcrowding.** Eggs spawned under natural conditions never have to face the overcrowded conditions of a beaker or petri dish. Minimize this by placing very few eggs in a single culture. Anything more than a thin layer on the bottom is too many.

4. **Keep Temperatures Constant.** Remember, marine animals are accustomed to very stable temperatures. Pay strict attention to the temperature requirements for the species you use! Moreover, slides heat up quickly on the microscope stage and culture solutions evaporate quickly, so do not expect to follow development for any length of time on a single slide. You will need to create "stock" solutions of embryos from which you can pipette out small aliquots for viewing.

OBTAINING GAMETES

Your instructor may perform the initial injections for you. Live sea urchins should be placed aboral side down in a shallow dish of sea water. Approximately 2 mL of 4% potassium chloride (KCl) should be injected into the oral side of the urchin (in the soft, fleshy tissue adjacent to the mouth) to induce release of gametes through the genital pores on the aboral side. It may take a few minutes for gametes to start "flowing." During this interim, place the sea urchin _oral side up_ over a dry petri dish until it begins to shed gametes into the dish (through pores on its aboral side). Occasionally a second injection of KCl is necessary to induce gamete release. If an urchin has not begun to shed gametes after 10 minutes, inject another 1–2 mL of KCl.

Your first task after injecting the urchins is to identify the gametes. The sex of the sea urchin can be determined by the color, texture and size of the gametes. The color will vary depending on the

species, so be sure you know what species you are using! In general, eggs are much larger and will quickly precipitate to the bottom of the beaker, while the tiny sperm cells will form a homogenous suspension in water.

1. **Male urchins**—leave aboral side down over a _dry_, sterile petri dish and allow the sperm to collect on the bottom of the dish. Cover this mixture when done to maintain sterility. Sperm may be dispensed later using _dry_ toothpicks.

2. **Female urchins**—suspend them over the opening of a beaker (aboral side down) so that their genital pores are in contact with the sterile sea water in the beaker. This will permit the eggs to drift to the bottom.

3. For an added measure of security after the eggs accumulate on the bottom, you may carefully decant the sea water, leaving the eggs in the bottom of the beaker, and replace the discarded water with new, sterile sea water. This step is particularly important if you wish to save the eggs.

FERTILIZATION

1. Use a sterile pipette to transfer a few eggs from the beaker to a clean glass depression slide. Holding the pipette vertically will allow the eggs to settle to the tip of the pipette to facilitate depositing on the slide.

2. Fertilize the eggs by dipping the end of a _dry_ toothpick into a petri dish containing sperm and swirling it gently through the sea water on the slide.

3. Cover with a coverslip.

4. Scan the slide quickly using low power to locate eggs with sperm swimming nearby.

5. Observe carefully and try to find eggs in the process of being fertilized.

NOTE: The actual fertilization event may occur too rapidly for you to observe. As soon as a sperm cell successfully penetrates the egg, you will see the thin fertilization membrane begin to lift away from the outer margin of the egg at the site where the sperm entered. This fertilization membrane will gradually envelope the entire egg in a protective barrier preventing other sperm from entering. Likewise, you can recognize eggs that have already been fertilized by the presence of this fertilization membrane already in place. You may also see many unsuccessful sperm attached to the sticky fertilization membrane trying vainly to deposit their payload in the egg.

6. If fertilization has not occurred within 5 minutes, try again using fresh sperm and eggs.

Your instructor will make a stock culture of fertilized eggs from which samples can be taken and viewed throughout the remainder of the laboratory period to monitor subsequent development.

CLEAVAGE

The rate of cleavage in the embryo will depend on the temperature and pH of the surrounding medium. In the sea urchin, cleavage is almost identical to that of the sea star until the fourth cleavage. At this point, sea urchin eggs begin to divide unevenly.

1. Thirty minutes after fertilization, examine another wet mount from the original container of fertilized eggs. Can you detect any changes in the cytoplasm of the egg?

2. If a sample of fertilized eggs from 3–5 hours ago is available, try to locate embryos in their fourth cleavage (i.e., 16-cell stage). Can you see the large and small cells?

3. At regular intervals throughout the remainder of the lab period, you should make wet mounts to determine whether subsequent cleavages have taken place. If you are very careful (and very lucky) you may actually see a cleavage event in progress. Be certain to keep accurate records of both time intervals and appearance of the embryo.

4. Your instructor will have cultures of fertilized eggs in various stages of development. Make wet mounts of some of these cultures, carefully noting the time of fertilization and calculating the age (in days and hours) of the culture at the time you examine it. What other developmental stages do you recognize?

5. Record the time (from fertilization) when each of these stages was first seen. Combine your results with data from other members of the class. Using these data you should be able to devise a developmental schedule for the species of sea urchin with which you have been working under the specific conditions in your lab.

Questions For Review

1. In your study of mitosis, why was the blastula used?

2. Interphase has sometimes been referred to as a "resting" phase. Explain why this term *inaccurately* describes interphase.

3. What are the major differences between mitosis and meiosis?

4. Define spermatogenesis and oogenesis.

5. Which of the following statements concerning gametogenesis is *false*?

   a. Gametogenesis results in the production of haploid cells.

   b. Spermatogenesis results in the production of 4 spermatozoa.

   c. Oogenesis results in the production of 4 ova.

   d. Both mitosis and meiosis are involved in gametogenesis.

6. Define the following terms:

   a. blastocoel _____

   _____

   b. archenteron _____

   _____

   c. blastomere _____

   _____

   d. blastula _____

   _____

   e. morula _____

   _____

   f. gastrula _____

   _____

   g. dorsal lip _____

   _____

   h. neural fold _____

   _____

7. Does the sea star embryo grow appreciably during embryonic development? Explain.

8. Are all of the organ systems fully developed in free-swimming larvae? Which organ systems are likely to develop first?

9. How does temperature affect embryonic development rates in sea urchins?

10. Would you expect this temperature-related pattern to be true in frogs as well? Explain.

Protista

After completing the exercises in this chapter, you should:

1. Be familiar with the distinguishing characteristics of all protists and the major subgroups within Protista.
2. Be familiar with the basic anatomy of representatives from many of the major taxa within each newly proposed kingdom.
3. Be able to recognize and distinguish members of each subgroup from one another.
4. Be able to discuss general physiological processes, such as feeding, reproduction, locomotion, and osmoregulation, in protists.
5. Be able to define all boldface terms.

The kingdom **Protista** was first proposed in 1969 by Robert Whitaker to include all unicellular, eukaryotic organisms. Subsequent research indicated that many multicellular organisms, traditionally classified as fungi or plants are actually more similar to protists than to true fungi or plants. Consequently, some of these groups, such as algae, slime molds and water molds, are now more properly classified as protists. Often the unicellular forms are informally referred to as **microscopic protists**, while the multicellular algae and slime molds are informally categorized as **macroscopic protists**. Today, the kingdom Protista has grown into an incredibly diverse assemblage of over 60,000 different known species. Despite their diversity, all protists are **eukaryotic**, meaning their cells contain membrane-bound nuclei and other membrane-bound organelles. In unicellular forms, all functions of life are performed within the limits of a single plasma membrane. There are no organs or tissues, but there is often division of labor within the cytoplasm. Specialized organelles are used for support, locomotion, reproduction, defense, osmoregulation, nutrient acquisition and even sensory systems. In colonial and multicellular forms, more complex regions approximating true tissues and organs have evolved to perform many of these functions.

Protists occur nearly everywhere on the earth where there is available water. They live in oceans, streams, lakes, puddles, damp soils, moist bark, underneath rocks, and even in the body fluids of plants and animals as symbionts or parasites. Some are plant-like in nature—sedentary, **autotrophic** organisms that make their own food through photosynthetic processes. Others are more animal-like—motile, **heterotrophic** organisms that ingest large food particles and digest them intracellularly. Others still are fungal-like in nature and absorb small organic compounds requiring no further breakdown. Many protists have flexible **plasma membranes**, while others possess rigid **cell walls** made of cellulose, calcium carbonate or silica. Some protists store their reserves of glucose as **starch** (as do plants) while others employ **lipids** or lipid by-products (as do animals). All protists can reproduce **asexually** by mitosis (generally by simple fission), yet many have the ability to reproduce **sexually** through a combination of meiosis and nuclear exchange. Many protists can form **cysts** which allow them to lie dormant for long periods of time to escape harsh environmental conditions that they would not survive in an active metabolic state.

Traditionally, biologists have used a five-kingdom system originally based on morphological similarities for classifying all living organisms (Monera, Protista, Fungi, Plantae, and Animalia). As research with protists continues, biologists are revising the traditional classification scheme of a single protist kingdom, based upon recent molecular data that more accurately describes the evolutionary history and relationships of the members of this kingdom. In fact, the original five-kingdom system is considered obsolete by many taxonomists. Its major fault lies in the discovery that protists do not represent a **monophyletic** group. Protists are **polyphyletic**, meaning the kingdom we call Protista actually includes members derived from two or more ancestral forms not common to all members. Put another way, all modern protists do not share the most recent common ancestor (Fig. 4.1). Instead, there are at least five distinct monophyletic lineages of organisms (proposed as separate kingdoms) that comprise modern protists. Thus we see that the term *protist* is probably best used only in

an informal sense to refer to unicellular eukaryotes and their close multicellular relatives.

Since many biology textbooks have adopted this newer classification scheme for protists, we have elected to follow this pattern of coverage for these groups in this manual. Whether or not this newer scheme remains, or is replaced in the future by yet another, even more detailed classification scheme, remains to be seen. As research using nucleic acid sequencing and detailed comparisons of cell structures progresses, biologists will continue to tease apart these complex relationships and provide us with an accurate classification system that reflects the true evolutionary history of these organisms.

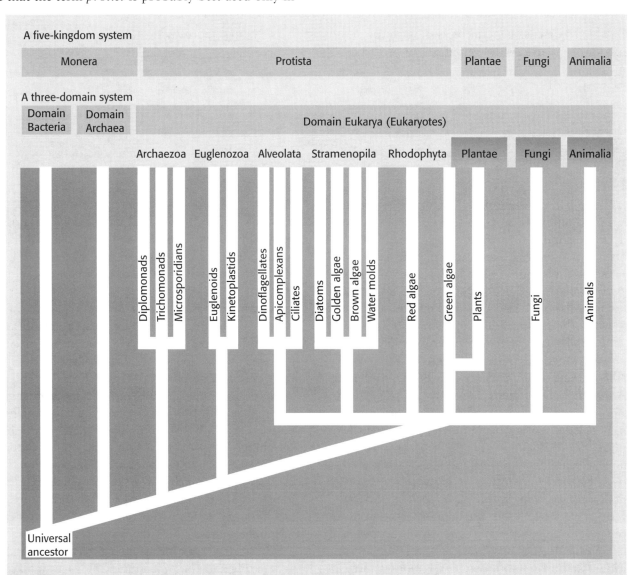

FIGURE 4.1 ▪ Revised phylogeny of eukaryotes with taxonomic categories of the traditional five-kingdom system and the newer three-domain system. Recent revisions to this phylogeny have resulted in a newly proposed classification scheme for the kingdom Protista, with advocates arguing for five separate protist kingdoms, each encompassing a single monophyletic lineage.

Euglenozoa (Euglenoids, kinetoplastids)

Materials needed:

- slide of *Euglena* w.m.
- live culture of *Euglena*
- slide of *Trypanosoma* w.m.
- blank slides
- coverslips
- disposable pipettes
- methylcellulose, Protoslo® or Detain™
- compound microscope

Though there are many protists with flagella, two groups of flagellates, the **euglenoids** and **kinetoplastids**, make up the monophyletic proposed kingdom Euglenozoa. All **Euglenozoans** are unicellular, motile flagellates that reproduce asexually by binary fission. Many species are heterotrophic, but some are also photosynthetic. All possess flexible cell membranes.

EUGLENOIDS (Ex. *Euglena*)

1. Obtain a prepared slide of a whole mount (w.m.) of *Euglena* and examine it using the medium or high power lens of your microscope.

2. *Euglena*, a complex unicellular organism with many visible organelles, typifies this subgroup. The characteristic feature of euglenoids is an anterior pocket that bears one or two **flagella** extending from the *anterior* end of the organism. Near the base of the primary flagellum is a pigmented **eyespot** that serves as a photodetector, providing chemical information to the cell about the intensity of light in its environment (Fig. 4.2). A large, central **nucleus** and many large **chloroplasts** should be evident, indicating that *Euglena* can photosynthesize. In fact the chloroplasts of euglenoids contain **chlorophyll b**, the same photosynthetic pigment used by all modern plants. In addition to autotrophic means of food production, *Euglena* can, when kept in dark environments, resort to heterotrophic methods of food acquisition—absorbing organic molecules from its environment. Although plant-like in their ability to photosynthesize, euglenoids are more animal-like in their method of food storage, storing glucose in the form of **lipids** rather than as starch.

3. Use a disposable pipette to place a drop of culture medium from a container of live *Euglena* on a clean microscope slide.

4. Add a coverslip and observe under medium or high power.

5. If the organisms are moving too quickly to observe easily, make another slide, this time adding a drop of methylcellulose, Protoslo® or Detain™ to slow their movement.

6. Notice the fluid, corkscrew motion that *Euglena* exhibits. Also notice in which direction the primary flagellum moves the organism. In the world of swimming flagellates, there are two types of flagella. One type, the *whiplash flagellum*, pushes the organism through the medium. A familiar example of this is a human sperm cell. The other type is the *tinsel flagellum*, which pulls the organism through its environment. Often a large, circular, **contractile vacuole** is evident near the anterior pocket. Its function is to pump out the excess water that continuously diffuses into the organism as a consequence of the osmotic gradient organisms living in a hypotonic (freshwater) environment must face.

7. Identify as many of the other organelles depicted in Figure 4.2 on a live *Euglena* as possible.

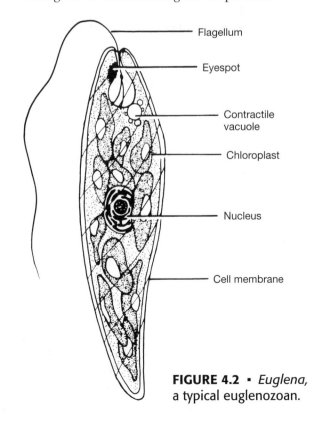

FIGURE 4.2 · *Euglena, a typical euglenozoan.*

Check Your Progress

1. Which type of flagellum, whiplash or tinsel, does *Euglena* possess?

2. What color is the photoreceptive eyespot of *Euglena*?

3. How might such an eyespot be advantageous for an autotrophic organism?

4. In what ways might *Euglena* be considered plant-like? In what ways is it animal-like?

KINETOPLASTIDS (Ex. *Trypanosoma*)

Kinetoplastids are unicellular, parasitic flagellates characterized by a single, large mitochondrion containing a **kinetoplast**—a unique organelle that houses extra-nuclear DNA. Many kinetoplastids, like *Trypanosoma*, are human pathogens. *Trypanosoma brucei* causes African sleeping sickness, a human disease that is spread by the bite of the tsetse fly. One reason this disease is still so widespread today in parts of Africa is that the molecular composition of these pathogens changes rapidly, preventing immunity from developing in hosts.

1. Obtain a w.m. slide of *Trypanosoma* and examine it using high power.

2. Notice that there are many trypanosomes in a single field of view, found among the red blood cells of their host. They reproduce in the blood, but do not harm the blood cells. Instead they absorb nutrients from the host's blood, multiply in the bloodstream and produce toxins which ultimately reach and affect the nervous system.

3. The trypanosome body is elongated and slightly twisted, bearing a distinct **nucleus** near the center (Fig. 4.3). A single **flagellum** runs closely along a fold of the plasma membrane, forming the **undulating membrane**, and extends beyond one end of the body. The **kinetoplast**, the characteristic feature of this group, is located at the end of the body opposite the flagellum, but is usually too small to been seen with a compound microscope.

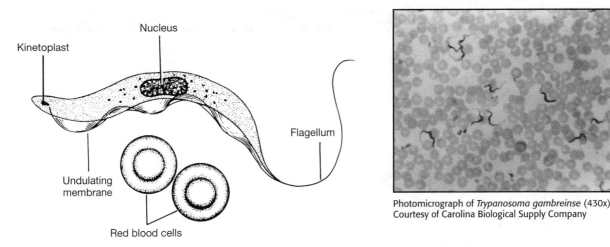

Photomicrograph of *Trypanosoma gambreinse* (430x)
Courtesy of Carolina Biological Supply Company

FIGURE 4.3 • Trypanosome parasites in blood—the causative agent of African sleeping sickness in humans.

Check Your Progress

1. Describe the length and width of a trypanosome compared to a red blood cell.

2. Are trypanosomes larger or smaller than *Euglena*?

3. Do trypanosomes possess chloroplasts? How do they obtain nutrients?

4. How are kinetoplastids similar to euglenoids?

Alveolata (Dinoflagellates, sporozoans, ciliates)

Materials needed:

- slide of dinoflagellates w.m.
- live culture of *Peridinium* or *Ceratium*
- slide of *Plasmodium* w.m.
- slide of *Paramecium* w.m.
- slide of *Paramecium* undergoing fission
- slide of *Paramecium* undergoing conjugation
- live culture of *Paramecium*
- live culture of *Vorticella*
- methylcellulose, Protoslo® or Detain™
- yeast mixture dyed with Congo Red
- blank slides
- coverslips
- disposable pipettes
- compound microscope

1. Obtain a prepared slide of dinoflagellates and examine with medium or high power.
2. Look for the two **perpendicular grooves** and the rigid **cellulose plates**. It may even be possible to make out **flagella** on well-prepared specimens.
3. Compare what your are observing with Figure 4.4.
4. Next make a wet mount of a living dinoflagellate (either *Peridinium* or *Ceratium*) and examine it under medium or high power.
5. If the organisms are moving too quickly to observe easily, make another slide, this time adding a drop of methylcellulose, Protoslo® or Detain™ to slow their movement.
6. Notice that, unlike euglenoids, dinoflagellates push themselves through the water with their flagella. Try to identify the two perpendicular grooves, the edges of the cellulose plates and the flagella on your live specimens.

Another monophyletic kingdom which has recently been proposed includes the dinoflagellates, sporozoans and ciliates and is called **Alveolata**. All members of this group of unicellular protists have small cavities under their cell surfaces called **alveoli**—the function of which is still unknown.

PYRROPHYTA (Dinoflagellates)

Dinoflagellates are mostly marine, unicellular autotrophs whose flagella arise from two **perpendicular grooves** formed along internal **cellulose plates**. One flagellum lies in a groove around the equator of the cell, and the other extends from the end of the cell in the other groove (Fig. 4.4). About 1,000 species of dinoflagellates have been described and they are among the most important primary photosynthetic producers of organic matter in the oceans. Their primary photosynthetic pigment is **chlorophyll c**. Reserves of glucose may be stored as **starch** or **lipids**. When nutrients are abundant, certain marine dinoflagellates experience large population "blooms" in such enormous numbers that they color the ocean waters a deep reddish hue for miles. The toxins and anoxic conditions produced by these "**red tides**" poison massive numbers of fish and make shellfish in these areas unfit for human consumption.

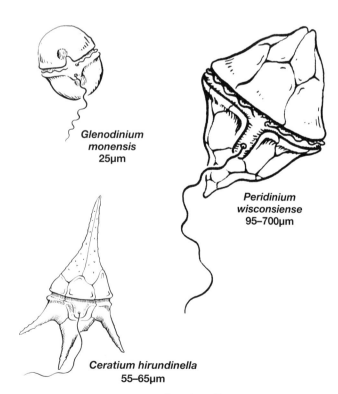

Glenodinium monensis 25µm

Peridinium wisconsiense 95–700µm

Ceratium hirundinella 55–65µm

FIGURE 4.4 ▪ Common dinoflagellates of the subgroup Pyrrophyta.

Check Your Progress

1. What ecological importance do dinoflagellates have?

2. What ecological problems do they occasionally cause?

3. Describe how the motion of dinoflagellates differs from that of euglenoids.

APICOMPLEXA (Sporozoans)

Members of the group **Apicomplexa** are unicellular parasites characterized by an apical complex of organelles that they use to penetrate the host cell. Sporozoans, as they are commonly called, infect nearly every major group of animals, from simple invertebrates to humans. So far around 3,900 different species have been identified—most infecting a different host! Though sporozoans are all nonmotile, they are highly specialized for their parasitic lifestyle. They rely on their hosts for both nutrients and dispersal and often have complex lifecycles that involve several host species.

1. Obtain a prepared slide of a blood smear containing *Plasmodium* and view it using the high power lens of your microscope. Like *Trypanosoma*, *Plasmodium* is a blood parasite carried by a secondary host, the female *Anopheles* mosquito, which transmits the parasite to humans through its saliva. The mosquito serves as a **vector** for the parasite, harboring the parasite and transmitting it, while remaining unaffected by the parasite. Humans, on the other hand, suffer gravely from the ensuing fever, chills, shaking and delirious fits that typify the unfortunate victims of malaria—a result of the toxins released by these parasites into the host's bloodstream.

2. Organisms should be visible on your slide *inside* each red blood cell as tiny black rings with small dots. Do not be confused by the larger, white blood cells with oddly shaped, purple nuclei.

These are normal constituents of blood. *Plasmodium* infects only the red blood cells. Notice too that, unlike *Trypanosoma*, *Plasmodium* actually enters the red blood cell. Depending upon the stage of infection depicted on your slide, you may observe one or more stages in the lifecycle of *Plasmodium* (Fig. 4.5).

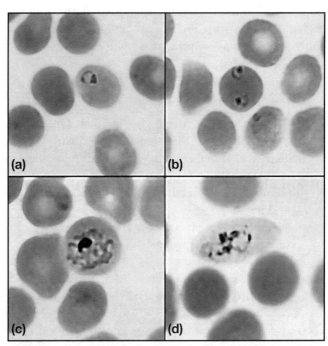

FIGURE 4.5 ▪ Several stages in the lifecycle of the sporozoan *Plasmodium falciparum*, a causative agent of malaria in humans. (a) The ring stage within a red blood cell, (b) a single red blood cell with a double infection, (c) a developing schizont, and (d) a gametocyte.

Check Your Progress

1. Do sporozoans (such as *Plasmodium*) possess organelles for locomotion?

2. Does every red blood cell contain a parasite?

3. Can you detect different stages of infection on the same slide?

CILIOPHORA (Ciliates)

More than 8,000 described species make up the subgroup **Ciliophora**. They are **unicellular heterotrophs** that possess **cilia** for locomotion and feeding, and have two types of nuclei: **micronuclei** and **macronuclei**. Most live as solitary organisms in fresh water, but their specific arrangements of cilia allow ciliates to be specialized for different lifestyles. Some are completely covered by cilia (i.e., *Paramecium*) whereas others have cilia clustered into a few rows along the body (i.e., *Vorticella*). Using their micronuclei, ciliates undergo an elaborate process of sexual gene shuffling called **conjugation**—a process we will examine in detail in *Paramecium*.

1. Obtain a prepared slide of a whole mount of *Paramecium* and examine under medium power.

2. Use Figure 4.6 to assist you in identifying the major organelles of this unicellular protist:

   pellicle–stiff outer covering that maintains basic cellular shape

   cilia–hair-like projections used for locomotion and feeding

   macronucleus–organelle containing many copies of a few genes; primarily controls metabolic processes of cell

   micronucleus–typical eukaryotic nucleus containing entire genome; essential for genetic recombination

   oral groove–lateral depression into which food is swept by ciliary currents

   cytopharynx–tubular invagination lined with cilia where food enters and food vacuoles form

   food vacuole–small, spherical organelle containing enzymes to digest food

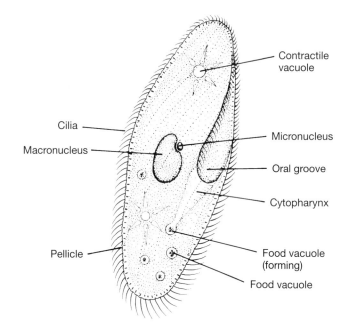

FIGURE 4.6 • Anatomy of *Paramecium caudatum*, a typical ciliate.

   contractile vacuole–one or more spherical organelles which pump out water to maintain the internal osmotic balance of the cell

3. Next obtain a prepared slide of *Paramecium* undergoing **binary fission** and examine under medium power. During this asexual process the micronucleus divides by **mitosis**, while the macronucleus and remainder of the original cell simply split in half to produce two genetically identical daughter cells. Notice the plane of division of the cell during fission.

4. Now obtain a prepared slide of conjugating *Paramecium* and examine under medium power. Your slide may show pairs of cells at several different stages of this process. During **conjugation**, two *Paramecium* align longitudinally and attach. The micronuclei in each cell undergo **meiosis** to create 4 haploid micronuclei per parent micronucleus (ciliates often have more than one micronucleus). The two cells simultaneously exchange some of these haploid micronuclei while the macronucleus and some of the micronuclei left in each original cell disintegrate. Next, the remaining and acquired micronuclei in each cell fuse to form a new, genetically "revised" micronucleus which divides mitotically to produce two daughter cells: a new macronucleus and a new micronucleus. Finally the two cells separate and usually undergo two more consecutive mitotic cell divisions to produce four daughter cells per parent cell.

5. Next make a wet mount from a culture of living *Paramecium*. Add a drop of methylcellulose, Protoslo® or Detain™ to the slide, place a coverslip on top and examine under medium power.

6. Identify as many of the structures depicted in Figure 4.6 as possible. If you watch carefully for a minute or two you should be able to see the contractile vacuoles in action as they fill with water and contract to pump the water out of the cell.

Check Your Progress

1. Give the function of each organelle listed below:

   a. oral groove —

   b. micronucleus —

   c. macronucleus —

   d. contractile vacuole —

   e. food vacuole —

   f. cilia —

2. Do the cilia of *Paramecium* beat in unison or in small groups?

3. Describe the process of feeding in *Paramecium*.

4. Is the plane of division during fission along the longitudinal or transverse axis?

5. Why is conjugation considered a form of sexual reproduction?

6. What evolutionary benefit does conjugation provide that fission does not?

In this next procedure you will examine *Vorticella*, a sessile, stalked ciliate commonly found in stagnant bodies of freshwater.

1. Prepare a wet mount from a culture of living *Vorticella*, add a coverslip and examine under medium or high power (Fig. 4.7).

2. Closing the iris diaphragm and adjusting the condenser on your microscope to increase image contrast usually improves observation of these small, transparent organisms. *Vorticella* typically remains attached to aquatic vegetation by a **contractile stalk** and possesses a funnel-shaped cell body with a ring of **cilia** around the larger, open end (**peristome**). The contractile body permits *Vorticella* to push its cell body further away from the substrate and neighboring individuals to compete for food. The cilia rapidly beat to create currents that pull food particles into the peristome.

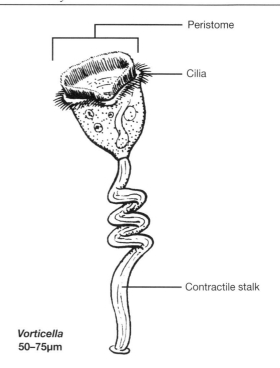

Vorticella
50–75μm

FIGURE 4.7 ▪ *Vorticella,* another common freshwater ciliate.

EXERCISE 4–3

Photographic Atlas Reference Page 22

Stramenopila (Diatoms, golden algae, brown algae and water molds)

Recent molecular data suggests that another seemingly diverse group of protists belong to a monophyletic lineage proposed as the kingdom **Stramenopila.** The distinguishing feature of the stramenopiles, as they are collectively called, is the presence of two flagella of unequal length, the longer of which bears rows of tubular hairs. The photosynthetic forms have unusual chloroplasts with two additional membranes, a small amount of cytoplasm and a vestigial nucleus, suggesting that these chloroplasts might have originally been eukaryotic endosymbionts. Even though many modern stramenopiles lack the characteristic pair of flagella, evidence suggests that they are descendants of ancestors that possessed the typical stramenopile flagella (much like modern whales are considered mammals even though they lack many of the characteristic features of most other modern mammals).

Materials needed:
- slide of mixed diatoms w.m.
- live culture of diatoms
- sample of diatomaceous earth
- dissecting needle
- dropper bottle of water
- live or preserved specimens of brown algae
- blank slides
- coverslips
- disposable pipettes
- compound microscope
- dissecting microscope

CHRYSOPHYTA OR BACILLARIOPHYTA (Diatoms)

The subgroup **Chrysophyta** (or Bacillariophyta) includes around 11,000 species of freshwater and marine unicellular autotrophs that have two-part cell walls made of silica, a natural glass. These two halves fit together like the top and bottom of a shoebox or petri dish. Locomotion occurs as diatoms slide along a thin film of cytoplasm secreted through tiny pores or slits in their cell walls. Diatom plastids contain a mixture of **chlorophyll c** and **fucoxanthin**, and many species store the glucose produced through photosynthesis as oils (**lipids**). Reproduction is most commonly asexual and some species are capable of forming resistant cysts.

Diatoms are a major component of the **plankton** that live in large numbers in the upper levels of the ocean. Even a bucket full of sea water may contain millions of microscopic diatoms! As a testament to their ecological importance (and abundance) over the course of life on earth, it is thought that much of the free oxygen in the atmosphere is a direct result of photosynthesis in marine diatoms! As these tiny organisms die, their hard siliceous shells settle to the ocean floor. Over time, these deposits can accumulate to massive depths approaching 300 feet! In fact, a large portion of our undersea oil reserves may be attributed to the accumulation of dead diatoms over hundreds of millions of years. In areas where ancient oceans have long since dried up, **diatomaceous earth** is mined and used commercially in a variety of products, including fine abrasives, silver polishes, toothpaste, filters, insulation and even a reflective roadway paint additive.

1. Obtain a prepared slide of mixed diatoms and examine it under medium or high magnification.

2. Notice the symmetry and delicate, ornamental patterns of the cell walls. Diatoms are categorized by their shape, which may be either pennate or centric. **Pennate** diatoms are rod-shaped with bilateral symmetry, while **centric** forms are circular, oval or elliptical and have radial symmetry (Fig. 4.8).

3. If the resolving power of your microscope is good, and you close the iris diaphragm and adjust the condenser on your microscope to increase the image contrast, you should be able to see the tiny grooves along the surface of the cell walls.

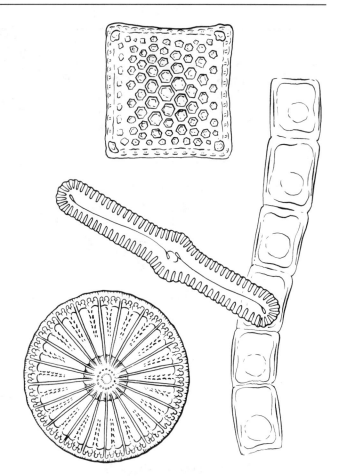

FIGURE 4.8 ▪ Examples of centric and pennate diatoms.

4. Examine a wet mount slide of a drop of culture medium containing live diatoms. Notice that most are golden in color. This is due to an excess of the pigment **fucoxanthin** which masks the green **chlorophyll c** also present.

5. Next, examine a sample of **diatomaceous earth**. Use a dissecting needle to scrape a small amount onto a clean slide and add a drop of water.

6. Stir well to dissolve before adding a coverslip, and observe under medium or high power. These are the skeletal remains of diatoms that lived millions of years ago.

Check Your Progress

1. Are your living diatoms pennate or centric?

2. Does your sample of diatomaceous earth contain more than one species?

3. Why do you think diatoms might be responsible for undersea oil reserves?

PHAEOPHYTA (Brown algae)

The brown algae consist of a group of around 1,500 mostly marine, multicellular autotrophs with cell walls made of **cellulose**. They are by far the largest and most complex protists. Some species, like the giant kelp, can reach lengths of up to 100 meters and can grow more than 60 meters in a single season, the fastest linear growth of any organism on the planet! Like diatoms, brown algae contain an abundance of the photosynthetic pigment **fucoxanthin**, giving them a brownish color, and store nutrients in the form of **lipids**. Brown algae also possess **chlorophyll a** and **chlorophyll c**, often in sufficient concentrations to mask their brown pigment and give them a greenish appearance.

In addition to their ecological role as a primary photosynthetic producer in oceans and intertidal zones, brown algae have considerable economic importance. Their cell walls contain **alginic acid**, a gummy polymer of sugar acids that is used as an emulsifier in ice cream, cosmetics and other products.

1. Examine the specimens of brown algae provided. Your instructor may provide specimens of *Laminaria*, *Fucus*, *Sargassum* or other species. Brown algae are composed of either branched filaments or leaf-like structures called **thalli** (singular = thallus).

2. The thallus is generally composed of three parts: an adhesive, root-like region called the **holdfast**, the elongated **stipe**, and flattened **blades** (Fig. 4.9). Though it is convenient to associate these three regions with the roots, stems and leaves of green plants, these three regions only superficially resemble plant structures and do not contain the many specialized internal tissues that true roots, stems and leaves possess.

3. Some species, like *Fucus* and *Sargassum*, possess small **air-filled bladders** along their blades.

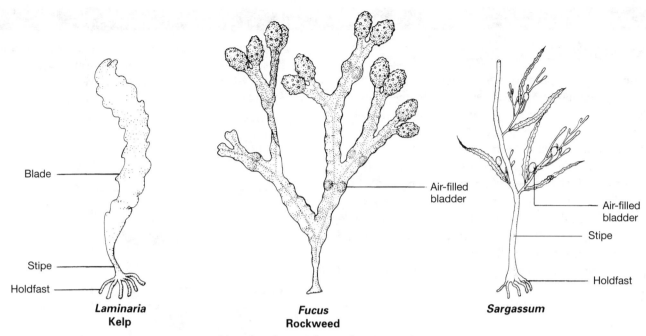

FIGURE 4.9 ▪ Three common species of brown algae (Phaeophyta).

Check Your Progress

1. How might the air-filled bladders of *Fucus* and *Sargassum* be advantageous to these organisms?

2. Name several features that brown algae share with diatoms.

3. In what ways are brown algae different than diatoms?

EXERCISE 4-4

Rhodophyta (Red algae)

The red algae comprise a single monophyletic group of approximately 4,000 species collectively called **Rhodophyta**. Their reddish hue is due to the presence of red and blue accessory pigments called **phycobilins** (phycoerythrin and phycocyanin). The amount of these pigments varies from species to species and also depends upon the depth at which the algae grow. As a result, many "red algae" are not very red at all, varying in color from almost black, to brown, to bright red, to even green. Since all red algae possess **chlorophyll d** as well, it is the relative proportions of phycoerythrin, phycocyanin and chlorophyll d which determine their outward color. The reddish phycoerythrin pigment allows red algae to live at much greater depths than their green or brown algae cousins, often as far as 100–200 meters below the surface. The red pigments (which appear red because they reflect red wavelengths) absorb the shorter, stronger blue and green wavelengths of light which penetrate deeper beneath the surface of the water.

All red algae are multicellular, macroscopic, autotrophic organisms with rigid cell walls made of cellulose. Like true plants, they store their reserves of glucose as **starch**. In addition to their unusual pigments, red algae lack flagellated gametes (unlike other algae) and must rely on slow amoeboid movements of these cells and water currents for their dispersal.

Red algae like *Porphyra* (a.k.a. nori) are a staple component of many Asian dishes, and extracts from other species provide the solidifying agent in **agar**, the universal laboratory medium on which bacteria and other microorganisms are grown. Another widespread commercial algal extract is **carrageenan**, a product used as a stabilizer in paints and cosmetics and as an emulsifying agent in ice cream, chocolate milk and salad dressings.

Materials needed:

- live or preserved specimens of *Polysiphonia* and *Porphyra*
- dissecting microscope

---

1. Examine the specimens of red algae provided. Use a dissecting scope for living material. Notice that in addition to color variations, different species of red algae vary in morphology. Some species, like *Polysiphonia*, are branched filaments, while others, like *Porphyra*, are thin, 2-cell layer sheets (Fig. 4.10).

(a) *Porphyra*

(b) *Ceramium*

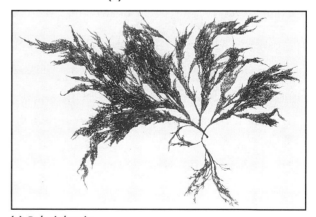

(c) *Polysiphonia*

FIGURE 4.10 ▪ Three common species of red algae (Rhodophyta).

Check Your Progress

1. What wavelengths of light would red algae be able to capture better than green algae?

2. Explain why this allows them to live at greater depths.

3. List several economic uses of red algae.

Chlorophyta (Green algae)

Though very diverse in shape and size, the green algae all share the characteristic bright green chloroplasts of modern plants. Molecular evidence suggests that modern plants and modern green algae shared a recent common evolutionary ancestor. Over 7,000 species of green algae have been identified, ranging from microscopic unicellular species, to tiny colonial forms, to large macroscopic, plant-like species. They have rigid cell walls made of **cellulose** and most live in fresh water environments, though many are marine, and some even manage an existence in moist, terrestrial habitats. Their characteristic grass-green color is due to an abundance of the photosynthetic pigment **chlorophyll b** which produces glucose that is stored in the form of **starch** in all green algae.

Because of the continuum of body forms present in modern green algae, this group provides an excellent model for determining the evolutionary origins of multicellular organisms. Think of the four examples of green algae we will observe as living representations of the evolutionary progression from single-celled life to complex, multicellular organisms.

Materials needed:
- live cultures of *Chlamydomonas* (+ and – strains)
- slide of *Spirogyra* undergoing conjugation
- live culture of *Spirogyra*
- slide of *Volvox* w.m.
- live culture of *Volvox*
- live or preserved specimens of *Ulva*
- blank slides
- depression slides
- coverslips
- methylcellulose, Protoslo® or Detain™
- toothpicks
- disposable pipettes
- compound microscope

UNICELLULAR GREEN ALGAE

The simplest green algae are unicellular algae, exemplified by *Chlamydomonas*. The tiny, egg-shaped, haploid cells of this species possess two flagella which they use to propel themselves in a whirling motion through their environment. *Chlamydomonas* probably resembles some of the earliest, most primitive green algae, both in its structure and its reproductive mode.

1. Prepare a wet mount by mixing a drop of culture medium containing live *Chlamydomonas* with a drop of methylcellulose, Protoslo® or Detain™.

2. Place a coverslip on the slide and examine using high power. Close the iris diaphragm and adjust the condenser of your microscope to increase contrast.

3. The first thing you probably notice about *Chlamydomonas* is their small size! Despite this feature, if you find a few stationary cells, you should be able to see their internal structures.

4. The most conspicuous organelles are the large **chloroplast** used for photosynthesis, and **pyrenoid** which functions in starch formation. It may be difficult to see the two **flagella** protruding from the tapered end, but you might be able to make out the photoreceptive **eyespot**, located near the base of the flagella (Fig. 4.11).

5. The method of sexual reproduction in *Chlamydomonas* is believed to have evolved early in the evolutionary lineage of green algae. In fact, cells of *Chlamydomonas* actually resemble gametes of more complex species of green algae. Because "male" and "female" sexes are similar in size and appearance, we classify the two compatible mating strains as plus (+) and minus (−). This condition is known as **isogamy**, so gametes that are morphologically indistinguishable are said to be **isogamous**. The opposite condition, in which male and female gametes visibly differ in size and shape, is known as **oogamy**. Humans provide a familiar example of a species with **oogamous** gametes.

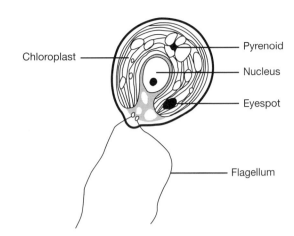

FIGURE 4.11 ▪ *Chlamydomonas*, a common unicellular green algal species.

6. To demonstrate sexual reproduction in *Chlamydomonas*, obtain a clean microscope slide and place a small drop of culture medium from the + strain next to a small drop of culture medium from the − strain.

7. Get a toothpick and coverslip and return to your microscope.

8. Focus on one of the droplets using medium power.

9. Then use the toothpick to gently mix the two drops together and examine the portion of the slide where the two strains meet.

10. Add the coverslip if you wish to use the high power lens.

11. You should be able to observe several distinct clumps of cells forming where the two strains came together. This clumping of + and − strains precedes the actual fusion of the two types of cells and their genetic material.

Check Your Progress

1. What other protist that you observed earlier had an eyespot? Was it also autotrophic?

FILAMENTOUS GREEN ALGAE

Filamentous forms of algae represent the next evolutionary step toward true multicellularity. If single cells divided, but remained attached by their cell walls on one side, and division occurred along the same plane, simple, unbranched filaments would form. One of the classic examples of this type of filamentous green algae is *Spirogyra*. Its name is derived from the unique spiral arrangement of its chloroplasts.

1. Prepare a wet mount with a few filaments of *Spirogyra* and examine using medium power. Notice that its simple filaments are one cell thick and unbranched.

2. Distinct **cell walls** are visible as are the spiral **chloroplasts** unique to this species of green algae. **Pyrenoids** may also be visible along the margins of the chloroplasts (Fig. 4.12). The **nucleus** is usually suspended from strands of cytoplasm near the center of the cell. Asexual reproduction occurs simply by **fragmentation** of individual filaments, while sexual reproduction occurs by a process known as **conjugation**.

3. Examine a prepared slide of *Spirogyra* undergoing conjugation. You may need to scan the slide carefully for several examples to see all stages of this process.

4. During conjugation, two compatible haploid filaments align and form cytoplasmic bridges, called **conjugation tubes**, that connect adjacent cells together. Next, the entire cytoplasmic contents of one cell coalesce into a "gamete" that migrates across this bridge to the adjacent cell and fuses with the other gamete forming a diploid **zygote** (Fig. 4.13). Later this diploid zygote will undergo meiosis to form four haploid nuclei. Three of the four nuclei die, leaving one functional haploid nucleus per cell.

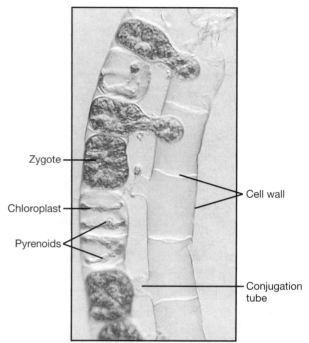

FIGURE 4.13 · Conjugation in filamentous green algae.

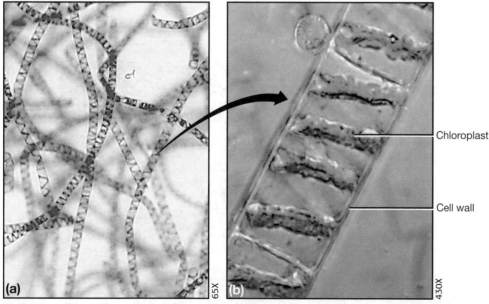

FIGURE 4.12 · Filaments of the green alga *Spirogyra* (Chlorophyta) and (b) highly magnified view of a single filament showing details of the cell structure.

Check Your Progress

1. Would you classify *Spirogyra* as a species with isogamous or oogamous gametes?

2. Explain why conjugation is considered a form of sexual reproduction.

COLONIAL GREEN ALGAE

Colonial forms of green algae are characterized by spherical aggregates of cells that have physiological connections between the cytoplasm of neighboring cells. Simple colonies may consist of as few as 4 to 8 cells, while complex colonies may contain thousands of cells. Even the most complex colonies demonstrate very little division of labor among cells within the colony—a marked distinction from true multicellular organisms. We will examine *Volvox*, one of the largest and most spectacular colonial green algae.

1. Obtain a prepared slide of *Volvox* and examine using low or medium power. *Volvox* colonies are hollow spheres consisting of anywhere from a few hundred to a few thousand small, biflagellated cells. The individual cells of *Volvox* resemble *Chlamydomonas*, the unicellular, biflagellated green alga you observed earlier. The tiny flagella beat in coordinated waves that cause *Volvox* to spin on their axes as they move through the water.

2. Using a depression slide, prepare a wet mount of *Volvox* from a live culture and examine under low or medium power. Notice the spinning motion they exhibit when they move.

3. You have probably also noticed that, in addition to the hundreds of **vegetative cells**, many colonies contain smaller spheres of cells within the larger colonies (Figure 4.14). These **daughter colonies** are produced by either sexual or asexual reproduction by the parent colony. During **asexual reproduction**, portions of the parent colony pinch inward forming daughter colonies that grow through mitotic cell divisions and are released when the parent colony ruptures. During **sexual reproduction**, several cells in the colony differentiate into motile sperm and others become nonmotile eggs. The sperm swim to the eggs and fuse with them to become zygotes. Zygotes develop into daughter colonies within the parent colony and are released when the parent colony breaks apart.

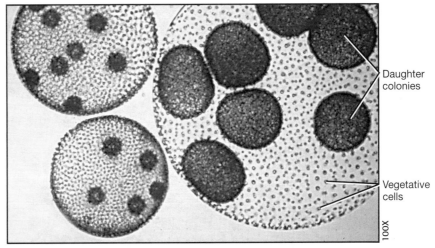

FIGURE 4.14 ▪ The colonial green alga, *Volvox*. Parent colonies are hollow spheres of hundreds to thousands of vegetative cells containing smaller, daughter colonies.

Check Your Progress

1. Based on the description of its gametes, is *Volvox* isogamous or oogamous?

2. Are daughter colonies the product of sexual reproduction? Are they the product of asexual reproduction?

MULTICELLULAR GREEN ALGAE

A final representative of the green algae illustrates the fourth level of cellular organization, those having a multicellular, tissue-like body. *Ulva*, commonly known as sea lettuce because of its green, leafy appearance, is one of the most common multicellular green algae (Figure 4.15). In addition to having an extremely complex body form for a protist, it displays the same elaborate reproductive cycle that all modern land plants possess.

1. Examine a living or preserved specimen of *Ulva*. *Ulva* represents another line of evolution among the green algae. The thallus of this species is flattened into a broad, leaf-like body that is only two cells thick. Some of the cells at the base of the thallus are modified into a holdfast which anchors the thallus in place against tidal currents.

FIGURE 4.15 ▪ The multicellular green alga, *Ulva*.

Check Your Progress

1. What other group of algae contain species with broad, flat leafy bodies only 2 cell layers thick?

Recurrent Body Forms (Amoebas, slime molds)

Amoebas and slime molds were previously classified as two distinct groups within the kingdom Protista, based on shared anatomical and behavioral similarities among organisms within each group. It now appears that amoeboid and slime mold body forms arose independently many times during the course of eukaryote evolution (Fig. 4.16). The anatomical and behavioral similarities among all amoebas (and among all slime molds) are due to convergent evolution rather than to a shared common ancestor. However, because both groups constitute a large and important sector of the protist world, coverage of them is essential to a broad understanding of the protists. Because molecular research has not yet provided a clear classification scheme for the members of these two groups, we will discuss only two common species and regard them as examples of recurrent body forms that have evolved independently along several distinct eukaryotic lineages.

> **Materials needed:**
> - slide of *Amoeba* w.m.
> - live culture of *Amoeba*
> - carmine in small bottle
> - dissecting needle
> - small squares of paper towel
> - dropper bottle of distilled water
> - live culture of slime mold (*Physarum*)
> - petri dish containing agar
> - small amount of crushed oatmeal
> - forceps
> - blank slides
> - coverslips
> - disposable pipettes
> - compound microscope
> - dissecting microscope

AMOEBAS

Amoebas are unicellular heterotrophs that lack cell walls. Despite their simplistic appearance, amoebas are quite specialized protists, tailor-made for their preferred habitats. Their methodical creeping motion and their manner of engulfing food particles equip them for life at the bottom of lakes and ponds amidst rich sources of sedentary organic particles. They have flexible plasma membranes that form **pseudopodia**—cytoplasmic extensions used for feeding and locomotion. Amoebas lack the ability to reproduce sexually. The majority are free-living, though some are parasitic, such as *Entamoeba histolytica*, the causative agent of amoebic dysentery in humans—a disease commonly spread through contaminated sources of drinking water or food. *Entamoeba histolytica* is one species of protist that can form a **cyst** by secreting a proteinaceous cell wall around itself and entering a state of dormancy. In the cyst stage, amoebas can survive prolonged periods of dryness and can even be dispersed by wind. If the cysts are ingested by a host, the cyst is activated and the amoeba crawls out and burrows into the host's intestinal wall, causing severe cramps and diarrhea, sometimes resulting in death by dehydration!

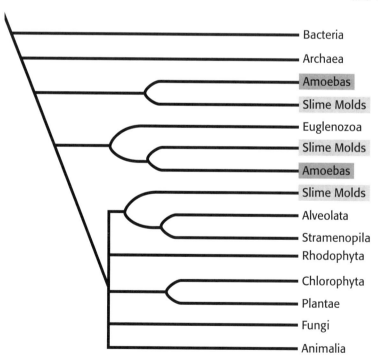

FIGURE 4.16 · Amoeboid and slime mold body forms have evolved repeatedly and independently along numerous protist lineages, creating challenges as to how to classify them taxonomically.

1. Obtain a prepared slide of amoebas and scan the slide using the low power objective to locate several stained amoebas.

2. While every specimen on your slide is likely captured in a different shape, many common features should be evident. A large, darkly stained, oval **nucleus** and numerous smaller, spherical **food vacuoles** should be visible (Fig. 4.17). Occasionally spherical **contractile vacuoles** will be present on prepared specimens. **Pseudopodia** should also be evident along the borders of the plasma membrane.

3. Now that you have a feel for what amoebas look like, obtain a drop of culture medium containing live amoebas and place on a slide with a coverslip.

4. If you are preparing the wet mount yourself, draw a small volume of culture from the bottom of the container using a pipette without stirring, bubbling or otherwise disturbing the culture medium. Though often difficult to find, amoebas typically crawl along the bottom of culture dishes and congregate near food sources or debris in the dish.

5. Scan your slide using low power to locate live amoebas. They will not be colored, like the stained specimens on the prepared slide you viewed earlier. Instead, they will be gray and semi-transparent with a granular appearance.

6. Once you've found an amoeba, move to medium power and close the iris diaphragm and adjust the condenser to increase the contrast (amoebas are nearly transparent!). There may be very little

outward movement initially, but upon closer examination under medium or high power magnification, **cytoplasmic streaming** should be visible.

7. Identify the following structures on your live amoeba: **nucleus, food vacuoles, contractile vacuole, pseudopodia**.

8. If you locate a clear, spherical contractile vacuole, watch it closely for a minute or two. You may see it contract as it pumps water out of the cell for the purpose of maintaining the proper internal ion concentrations inside the amoeboid cell.

9. Amoebas engulf their food by a process called **phagocytosis** in which extensions of the plasma membrane (pseudopodia) elongate and surround the food item. The portion of plasma membrane enclosing the food particle pinches off and forms a food vacuole inside the amoeba where digestive enzymes begin breaking down the food.

10. To demonstrate this process, place a drop of distilled water against one edge of the coverslip nearest the amoeba.

11. Use the tip of a dissecting needle to pick up a few carmine crystals and deposit them into the water droplet.

12. Hold a small piece of paper towel against the other side of the coverslip, *opposite* the water droplet containing the carmine, and draw the suspension beneath the coverslip.

13. Observe the amoeba and you may be lucky enough to catch it in the act of feeding!

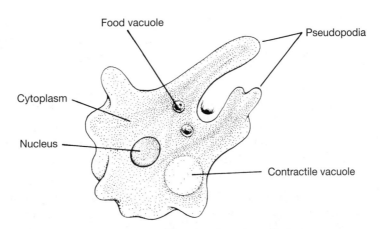

FIGURE 4.17 • Anatomy of *Amoeba proteus,* a representative amoeba.

Check Your Progress

1. What is phagocytosis? What function does it serve?

2. Relative to other unicellular protists, how would you characterize the size of amoebas?

3. Why would excess water tend to accumulate in amoebas?

4. Do amoebas have a rigid cell wall? How can you tell?

5. Do amoebas have a "permanent" anterior end like many other motile protists?

SLIME MOLDS

Slime molds represent a truly unique collection of organisms—one that has given taxonomists and evolutionary biologists difficulties for decades. Once thought to be fungi, slime molds only superficially resemble members of this kingdom, lacking the prominent hyphae and chitinous cell walls that all fungi possess. The three groups of slime molds seem so similar that, until recently, biologists grouped them in a single phylum within the kingdom Protista. Now recent evidence has shown that the three groups of slime molds are actually so different that they should be considered as separate kingdoms! Slime molds share only a few general characteristics. They are all **motile, absorptive heterotrophs** that lack rigid cell walls and form spores on erect **fruiting bodies**. During the course of their lifecycle, slime molds undergo dramatic changes in organization. They may be found as dormant, isolated clusters of **sporangiophores** or expansive, web-like networks of brilliantly-colored protoplasm. Favoring cool, moist and dark habitats, slime molds are most commonly found in forests, particularly on the undersides of decaying bark or logs.

Slime molds come in two basic body forms: **cellular** and **plasmodial** (the latter often referred to as acellular). Approximately 500 species of plasmodial slime molds have been described, while only around 65 species of cellular slime molds are known.

Cellular slime molds (former phylum Acrasiomycota) consist of large numbers of amoeboid cells, called **myxamoebas**, which have single haploid nuclei. This collective mass of individual cells creeps over the substrate engulfing food particles by **phago-cytosis**. This lifecycle stage may persist indefinitely as long as food is available. When food becomes scarce or conditions become otherwise unfavorable for growth, the individual cells of a cellular slime mold coalesce to form delicate, spore-bearing, fruiting bodies called **sporangiophores** which release spores. These spores settle and, when conditions are favorable again, germinate into new myxamoebas to continue the lifecycle.

Plasmodial slime molds (former phylum Myxomycota) form an acellular, wall-less mass of cytoplasm with numerous diploid nuclei. This vegetative mass, called the **plasmodium**, streams very slowly over the substrate, expanding into a beautiful and fragile network of fine, cytoplasmic filaments enclosed by a single plasma membrane (Figure 4.18). While the outer surface of the membrane is drier and normally less fluid, the inner cytoplasm remains very fluid, pulsating back and forth along this network of "veins" by **cytoplasmic streaming**. Plasmodial slime molds move by extending elongated regions of their plasma membrane outward as cytoplasm rushes into the new region and away from the older one. As it moves, the plasmodium engulfs food particles such as bacteria, yeasts, fungal spores, other microorganisms and decaying organic matter. In this manner they may grow indefinitely, provided resources remain available. As with their cellular counterpart, if conditions for growth become unfavorable, plasmodial slime molds can form **sporangiophores**. Alternatively, plasmodial slime molds may form an irregular, resistant, hardened mass called a **sclerotium**. This is a resting stage which rapidly becomes active again once favorable environmental conditions are restored.

1. Using your dissecting microscopes, examine a culture of *Physarum* in the plasmodial stage growing on oatmeal flakes.

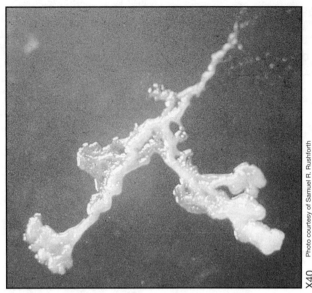

FIGURE 4.18 ▪ The plasmoidal slime mold, *Physarum*.

Photo courtesy of Samuel R. Rushforth

X40

2. Use the highest magnification available and shine the light from beneath the specimen to see **cytoplasmic streaming**. This process helps distribute nutrients and oxygen along the length of the veins, as well as allows for growth and movement of the plasmodium.

3. If you cannot see any cytoplasmic movement with your dissecting scope, your instructor may allow you to remove a small piece of the plasmodium and place it on a clean microscope slide to examine under your compound microscope's low or medium power lens.

Check Your Progress

1. Describe the color and texture of the plasmodium of *Physarum*.

2. Does the cytoplasm always move in one particular direction?

> **NOTE:** Check with your instructor to see if you will be completing the following exercise and whether you will be working in groups with other students.

OBSERVING SLIME MOLD GROWTH

1. Obtain a petri dish containing agar and a small piece of filter paper containing the resting **sclerotium** phase of a slime mold.

2. Use forceps to gently place the piece of filter paper in the center of the petri dish (Fig. 4.19).

3. Sprinkle the filter paper with a few crushed oatmeal flakes.

4. Moisten the oatmeal with a few drops of water.

5. Cover the dish and set in a dark place.

6. Examine 24 hours later for growth and again during next week's lab.

7. Record your observations in the spaces provided.

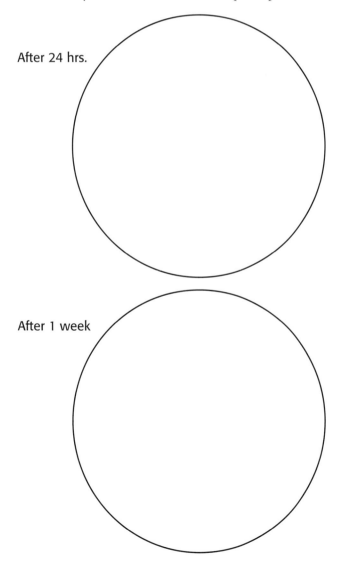

After 24 hrs.

After 1 week

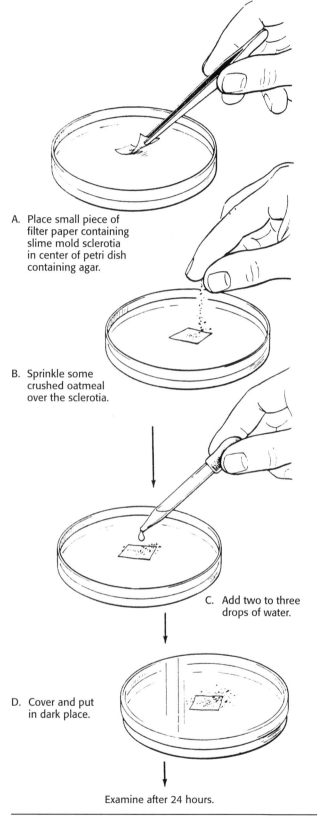

A. Place small piece of filter paper containing slime mold sclerotia in center of petri dish containing agar.

B. Sprinkle some crushed oatmeal over the sclerotia.

C. Add two to three drops of water.

D. Cover and put in dark place.

Examine after 24 hours.

From *Laboratory Outlines in Biology, VI* by Peter Abramoff and Robert G. Thomson 1994, 1991, 1986, 1982, 1972, 1966 , 1964 by W. H. Freeman and Company. Used with permission.

FIGURE 4.19 · Procedures for growing slime molds on enriched agar.

Questions For Review

1. Name three general methods of locomotion exhibited by protists and examples of organisms that utilize each method.

   Method of Locomotion   **Example**

   1.

   2.

   3.

2. What structures or features do all protists share in common?

3. Match the kingdom with its distinguishing feature.

   _____ a. Euglenozoa   1. two unequal flagella, the longer one with hairs

   _____ b. Alveolata   2. have chlorophyll b; store glucose as starch

   _____ c. Stramenopila   3. monophyletic group of unicellular flagellates

   _____ d. Rhodophyta   4. have small cavities under cell surface

   _____ e. Chlorophyta   5. have phycoerythrin, phycocyanin and chlorophyll d

4. Match the group with its correct description or characteristic.

   _____ a. kinetoplastids   1. autotrophs with two-part shells made of silica

   _____ b. euglenoids   2. unicellular heterotrophs with two types of nuclei

   _____ c. diatoms   3. multicellular autotrophs with fucoxanthin

   _____ d. ciliates   4. nonmotile parasites with penetrating apical region

   _____ e. dinoflagellates   5. flagella arise from anterior pocket

   _____ f. sporozoans   6. flagella arise from grooves in cellulose plates

   _____ g. red algae   7. multicellular autotrophs with phycobilins

   _____ h. brown algae   8. unicellular parasites with one large mitochondrion

5. Are the holdfast, stipe and blades of brown algae the same as roots, stems and leaves of land plants? Why or why not?

6. If brown algae and red algae both contain chlorophyll why do they not appear green?

7. What four levels of cellular organization (or body forms) are present within the subgroup Chlorophyta? Rank your answers in order of increasing complexity.

   1.

   2.

   3.

   4.

8. Give definitions that distinguish between the terms isogamous and oogamous.

9. What is the difference between the plasmodium and sclerotium of a slime mold?

Porifera

After completing the exercises in this chapter, you should:

1. Be able to recognize and describe the cellular organization of sponges.
2. Recognize the skeletal elements of sponges and be able to describe their composition.
3. Understand the pattern of water flow through the sponge and its significance in feeding and reproduction.
4. Be able to approximate the flow rate of a small freshwater sponge.
5. Understand the process of reaggregation in sponges.
6. Be able to define all boldface terms.

EXERCISE 5–1	**Photographic Atlas Reference Pages 29–32**

Sponge Anatomy

> *Materials needed:*
> • slides of *Grantia* (or *Scypha*)—c.s. and l.s.
> • compound microscope

The phylum Porifera includes around 10,000 species of sponges. Despite this seemingly large number, all sponges share many anatomical similarities which tie in to their common lifestyle. Sponges are sedentary animals that lack tissues or organs—characteristics that cause many zoologists to classify them as an outgroup to the evolutionary line that leads to all other modern animals. Their "bodies" consist of 4 primary cell types arranged in loose aggregates around a system of pores and canals through which sponges pass water. While most species are marine, there are around 150 freshwater species that have been described.

1. Obtain a prepared slide of a longitudinal-section through *Grantia, Scypha* or another sycon sponge. As you view the slide, try to visualize the path of water flowing from the outside of the sponge toward the inside. Water flow is *crucial* to understanding the anatomy of sponges and the reasons for their cellular organization. Since sponges do not move, every aspect of their existence, from feeding to reproduction, hinges on water moving through them.

2. The most obvious landmark on your slide is the large central cavity called the **spongocoel** that passes through the center of the sponge (Fig. 5.1). Before reaching the spongocoel, water passes through a complex series of chambers and canals in the sponge and, in doing so, is stripped of most food particles and oxygen it contains.

3. Water enters the sponge through pores on the body surface called **ostia**. Ostia channel water down the **incurrent canals** to a larger number of tiny pores scattered along the folds of the incurrent canals (Fig. 5.1).

4. These tiny pores are called **prosopyles**. The prosopyles channel water into flagellated chambers known as **radial canals**. Specialized cells called **choanocytes** lining the interior surface of the radial canals trap small food particles with their flagella and engulf them through phagocytosis. The beating of the choanocytes' flagella actually creates the water currents that flow through the sponge body.

5. Water then passes out of each radial canal though a larger opening called the **apopyle** into the **spongocoel** where it collects (Fig. 5.1b).

6. Finally the water is pushed out of the large **osculum** which is usually located at the top of the sponge. One of the ways sponges reproduce is by shedding gametes (much like the sea urchins you studied in the previous chapter). When water flows out of the osculum, gametes may be carried out and collected by nearby sponges as water flows into them.

7. Now examine a prepared slide of a cross-section through this same species of sponge. Identify all of the regions and structures previously discussed, keeping in mind that the orientation of this section is perpendicular to the longitudinal section viewed earlier.

8. Use Figure 5.2 to assist you in identification of the structures and regions on the cross-section.

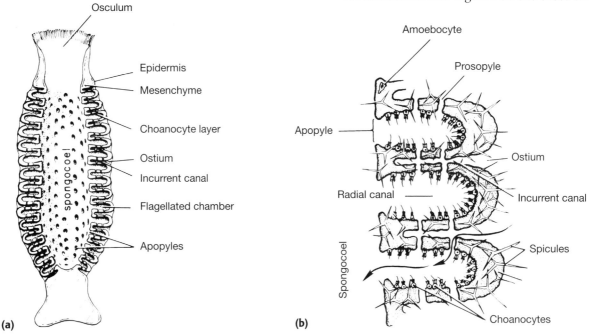

(a) (b)

FIGURE 5.1 ▪ Longitudinal-sections through (a) a simple sponge and (b) close-up depicting organization of the canals and chambers comprising the body wall of a simple sponge.

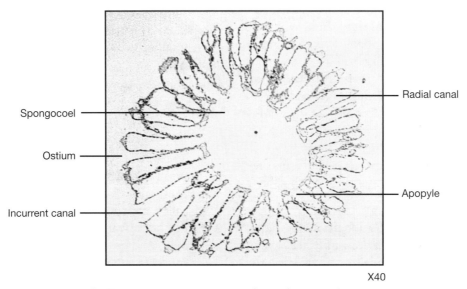

X40

FIGURE 5.2 ▪ Cross-section through a simple sponge.

Check Your Progress

1. On the diagram of the sponge below (Fig. 5.3) label the following regions: ostium, apopyle, prosopyle, spongocoel and radial canals.

2. On this same diagram, draw arrows to represent the flow of water through the canals and chambers of the sponge.

3. On this same diagram indicate the location in the sponge body where food particles are extracted from the water.

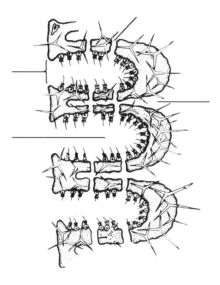

FIGURE 5.3 ▪ Unlabeled diagram of a simple sponge.

EXERCISE 5–2   **Photographic Atlas Reference Pages 31–32**

Observation of Spicules

Materials needed:
- preserved or living sponge
- bleach
- glass depression slides
- coverslips
- compound microscope

One of the characteristics zoologists use to classify sponges into taxonomic groups is spicule structure. Spicules (and spongin) are the skeletal elements of sponges and are secreted by **amoebocytes,** mobile cells specialized for distributing food throughout the sponge and for producing its skeleton. In many sponges this skeleton is comprised of hard, crystalline **spicules** formed from either calcium carbonate or silicon. Other sponges, like the exotic bath sponges that were formerly popular consumer items, secrete a proteinaceous material called **spongin** that is more flexible. Spicules come in many shapes and sizes and may have a haphazard arrangement within the sponge or be intricately woven to form dazzling geometric patterns. One of the most beautiful and delicate examples of this phenomenon is the intricate lattice of silicon filaments (i.e., glass) that the Venus Flower Basket (*Euplectella sp.*) forms.

1. Obtain a small piece of living or preserved sponge (*Grantia* will work fine for this exercise).

2. Place a drop of water in the depression well of the slide.

3. Use your coverslip to gently crush a small piece of sponge *on the edge of the depression slide* and scrape this crushed preparation into the water.

4. Add 1–2 drops of bleach to this mixture to dissolve the protein and other tissues, thereby exposing the spicules.

5. Cover with the coverslip.

6. Examine this preparation under low power, then higher magnification, to view the spicules.

7. Sketch several of the spicules in the space below.

EXERCISE 5–3   **Photographic Atlas Reference Pages 29–32**

Calculating Flow Rate in Freshwater Sponges

> *Materials needed:*
> - small freshwater sponges
> - clear petri dishes
> - pond water
> - clear plastic rulers/ocular micrometers
> - powdered algae (*Spirulina*)
> - small metal spatula
> - video microscope *or* camcorder and dissecting scopes

Although typically much smaller in size than their marine cousins, freshwater sponges play an important ecological role in their environment. Their apparent lack of behavior is deceiving, for these simple creatures are constantly pulling water through their porous bodies to extract suspended food particles and oxygen. While this is a necessary process for sponges to feed and breathe, it has the secondary effect of creating small "micro-currents" in the water surrounding sponges which keeps the water from becoming stagnant and contributes to the natural filtration process of lakes and ponds. These micro-currents undoubtedly play an important role in freshwater systems since sponges keep moving "new" water into their immediate vicinity and pushing "old," oxygen-depleted water away. But how could a sponge the size of a postage stamp have any dramatic effect on the massive volume of water surrounding it? In this next exercise, you will employ a simple but fascinating method for approximating the volume of water that a sponge can "filter" through its body in a given time frame. You may be surprised at the staggering volume of water that such a small sponge can move over the course of a day!

1. Obtain a small freshwater sponge and submerge it in a clear petri dish filled with pond water. The shape does not matter a great deal as long as one or more oscula are visible. Ideally your sponge should have one distinct osculum, preferably accentuated by a thin, transparent, chimney-like tube projecting from the osculum.

2. Observe the sponge with a dissecting microscope to locate the osculum.

3. Measure the diameter of the osculum as precisely as possible. Estimate to 1–2 decimal places if necessary.

   Osculum diameter = _____ mm

4. Calculate the cross-sectional area of the osculum. Area = πr^2; where r = radius and π = 3.142. Remember the radius is one-half the diameter.

   Cross-sectional area = _____ mm²

5. Rotate the sponge so that the osculum is pointing sideways in the dish (perpendicular to the plane of view of the microscope).

6. Adjust the level of water in the petri dish so that the sponge is submerged approximately 1 cm below the water surface.

7. To calculate the flow rate and resultant water volume processed by the sponge, you will need some type of video recording system. Your instructor will go over the details of the specific video recording setup you will use. See Figure 5.4 for complete experimental setup.

8. Position the petri dish containing the sponge in the camera's field of view and focus the lens on the osculum.

9. Adjust the shutter speed of the camera to 1/250th or 1/500th seconds and insert a blank videotape in the recorder.

10. Using a small spatula, lightly dust the surface of the water with a tiny amount of powdered algae (*Spirulina*).

11. Gently stir the water to suspend the algal particles in the solution. The algal particles will vary in size and some of the larger particles will be visible as they "shoot" out of the osculum.

12. Adjust the focal plane of the camera so that the algal particles leaving the center of the osculum are in focus. Make sure that the particles leaving the sponge are visible before proceeding.

13. Begin recording to make sure everything is setup properly, then hit "PAUSE" on the recording device.

14. Carefully insert a section of a transparent ruler mounted on a small "handle" into the petri dish next to the osculum (or use an ocular micrometer on the microscope).

15. Resume recording and shoot an additional 5–10 seconds of footage.

16. Rewind the videotape and review the last 5–10 seconds of footage frame by frame.

17. You will need to find two adjacent frames on the tape that have the same particle in view, leaving the osculum. Measure the distance that the particle traveled.

    Particle distance = _____ mm

18. Multiply the cross-sectional area of the osculum (calculated in Step #4) by the distance the particle traveled in the two adjacent video frames (calculated in Step #17) and record the product below. This number approximates the volume of the cylinder of water that the sponge extruded in the time that elapsed between the two video frames.

    Volume of water = _____ mm³ (or mL)

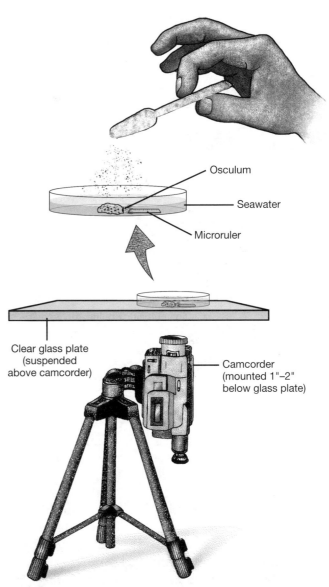

FIGURE 5.4 · Diagram of video recording setup for monitoring water flow in sponges.

19. The standard frame speed of a video recorder is 1/30th of a second. So the volume of water calculated above represents the amount of water the sponge can extrude in 1/30th of a second. Multiply this value by 1800 to estimate the volume of water the sponge would filter in 1 minute (the flow rate of the sponge).

Flow rate = _____ mL/minute

Check Your Progress

1. Assuming that a sponge has a constant flow rate over the course of a day, how many mL of water can this sponge filter in 24 hours?

2. How many liters of water would this represent?

EXERCISE 5–4

Sponge Reaggregation

> *Materials needed:*
> - living marine sponges
> - Ca^{2+}/Mg^{2+} free sea water
> - normal sea water
> - silk blotting cloth or cheese cloth
> - disposable pipettes
> - blank slides
> - coverslips
> - 25 mL or 50 mL beakers
> - compound microscope

One of the most fascinating properties of sponges which illustrates the loose physiological connections between their cells is their remarkable ability to reaggregate. If the cellular connections that hold the sponge cells together are broken, the sponge will fall apart into individual cells. However, these cells can spontaneously reassemble to form a functional sponge again. This feat cannot be accomplished by "higher" organisms with true tissues; the physiological connections between their cells are too permanent to permit disruption—even for a brief time. In this experiment you will observe the reaggregation phenomenon and will test the specificity of this process. Two species of sponges (probably *Microciona prolifera* and *Haliclona occulata*) have been soaking in Ca^{2+}/Mg^{2+} free sea water overnight. Calcium and magnesium are necessary elements for intercellular adhesion, therefore deprivation of these ions facilitates dissociation of the cells.

1. Prepare individual suspensions of dissociated cells of these two species by pressing a small piece of each sponge through a piece of silk blotting cloth (or cheese cloth) directly into a beaker containing 25 mL of normal sea water.

2. Since *Microciona prolifera* is bright red and *Haliclona occulata* is purple, you will be able to tell the two suspensions of dissociated cells apart. One will appear pink and the other blue.

3. Using a pipette, remove a sample from one of the beakers.

4. Place a drop on a slide, cover it and examine it with a microscope. Can you distinguish the individual cells?

5. Remove 1.0 mL from each beaker and mix the two samples in a small beaker containing normal sea water.

6. After 30, 60 and 90 minutes, remove small volumes of the suspension and make a wet mount. Look for thin protoplasmic extensions (**filopodia**) put out by the small clumps of aggregating cells.

Can you determine any differences in the size of aggregates? Are the aggregates formed from a combination of the red and purple cells or is there a segregation of cell types?

7. Record your observations in Table 5.1 and sketch a few cell clumps in the space below.

TABLE 5.1 ▪ Reaggregation of Sponges in Normal Sea Water

	Temperature		
	30 Minutes	**60 Minutes**	**90 Minutes**
Observations of sponge reaggregation			
Drawings of aggregates			

Questions For Review

1. What is the advantage of a folded or convoluted body wall in sponges?

2. Why are sponges placed in a taxonomic group beside the other animals and considered an evolutionary dead end by many zoologists?

3. What cell type in sponges is responsible for producing the water current through the sponge? What other important function do these cells serve?

4. Discuss the ecological benefits that the presence of sponges may have on aquatic systems.

5. Why were living sponges placed in Ca^{2+}/Mg^{2+} free sea water overnight? Hypothesize about the outcome of that exercise if this step had been omitted.

Cnidaria

6

After completing the exercises in this chapter, you should:

1. Be familiar with the basic anatomy of the representatives from the three major classes of cnidarians.

2. Be able to distinguish members of each class from one another.

3. Understand the difference between the polyp and the medusa body forms.

4. Understand the meaning of diploblastic tissue organization.

5. Be able to categorize and describe the proximate mechanisms of feeding behavior in cnidarians.

6. Be able to define all boldface terms.

The phylum Cnidaria is composed of three major classes of mostly marine carnivores: Hydrozoa (*Hydra*, *Obelia*, Portuguese man-o-war), Scyphozoa (true jellyfish) and Anthozoa (sea anemones and colonial corals). Of the 10,000 aquatic species estimated to exist in this phylum, the majority are marine. All cnidarians are **radially symmetrical** and most are **metamorphic**, meaning they have different body forms during their lives. The **polyp** form is generally represented by a cylindrical organism which remains attached to the substrate by a short stalk. The **medusa** is a more circular, free-swimming form resembling the familiar jellyfish in morphology (Fig. 6.1). The classes of cnidarians are distinguished primarily by the dominance of the polyp or medusa stage in the lifecycle. In hydrozoans the polyp form predominates the lifecycle, while in scyphozoans it is the medusa form which predominates. Anthozoans

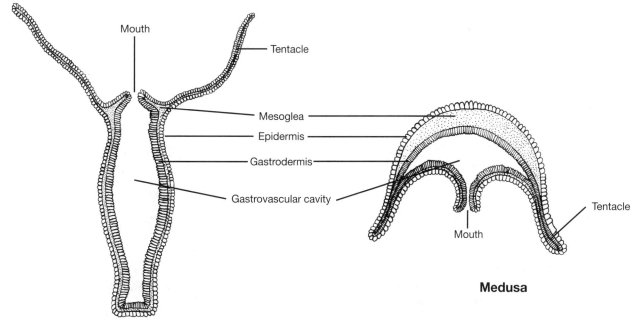

Mouth

Tentacle

Mesoglea

Epidermis

Gastrodermis

Gastrovascular cavity

Mouth

Tentacle

Medusa

Polyp

FIGURE 6.1 ▪ Body forms of Cnidaria.

exist only as polyps; the medusa stage has been completely lost. Another universal characteristic of this phylum is the presence of tentacles armed with stinging cells used for defense and for capturing food. These cells are called **cnidocytes**, from which term the phylum derives its name. Unlike sponges, the cnidarian body is arranged into two discrete tissue layers, giving them a **diploblastic** arrangement.

There is an outer **epidermis** and an inner **gastrodermis**. Sandwiched between these two cellular layers is an inert, gelatinous substance called **mesoglea**. Thus cnidarians possess a more complex organization of their cells than sponges and are consequently capable of many more intricate behaviors as a result of the specializations of their two cell layers.

EXERCISE 6–1   **Photographic Atlas Reference Pages 33–36**

Hydrozoan Anatomy

Most members of this class undergo a regular alternation of generations between body forms that utilize asexual reproduction (polyp form) and sexual reproduction (medusa form). In general the polyp is the predominant body form, and in many groups polyps are assembled into colonies, as in the case of *Obelia*. Members of the genus *Hydra* are unusual in that they do not produce medusae. Rather *Hydra* exist as single, mobile polyps that reproduce either sexually (through production of sperm and eggs) or asexually (through budding). In either scenario a new polyp is the result.

> *Materials needed:*
> - slides of *Hydra* w.m., c.s. and l.s.
> - slide of *Obelia* polyp w.m.
> - slide of *Obelia* medusa w.m.
> - compound microscope

SOLITARY HYDROZOANS

1. Obtain a prepared slide of a whole mount (w.m.) of *Hydra littoralis*. These small hydrozoans exist as polyps in shallow, freshwater ponds and streams. Their diminutive size is deceptive, for they are fierce predators of small aquatic invertebrates as they sit motionless among the submerged rocks, twigs and vegetation with their tentacles outstretched waiting for prey to pass too closely. Using a microscope, examine the w.m. slide and locate the following structures: **tentacles, hypostome, basal disc, bud** (may not be present), **gonads** (ovaries or testes). Anatomical structures are represented in Figure 6.2 and defined in Table 6.1.

2. If available, examine a longitudinal section (l.s.) through *Hydra*. In addition to the structures visible on the whole mount, locate the following: **mouth, gastrovascular cavity** (coelenteron), **epidermis, gastrodermis, mesoglea**.

3. Finally, examine a cross-section (c.s.) through *Hydra* and locate the following structures: **cnidocytes, gastrodermis, epidermis, mesoglea, gastrovascular cavity**.

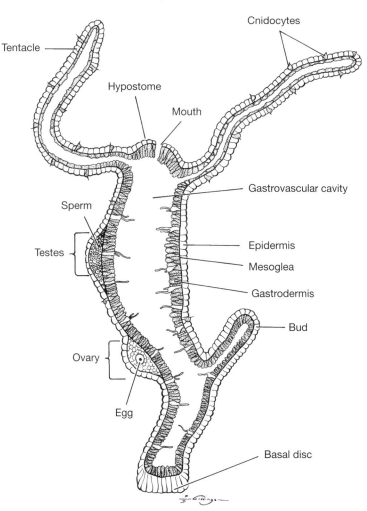

FIGURE 6.2 • Longitudinal section through *Hydra*.

TABLE 6.1 ▪ Anatomy of *Hydra*

Structure	Function
Tentacles	Defense and prey capture
Mouth	Ingestion of food and elimination of indigestible particles (egestion)
Hypostome	Elevated mound of tissue which expands or contracts to regulate size of mouth opening
Cnidocytes	Specialized stinging cells located in the epidermal layer of the tentacles
Gastrovascular cavity (coelenteron)	Chamber within which extracellular digestion of prey occurs; only opening is through the mouth
Bud	Product of asexual reproduction; will fall off when mature and become a self-sufficient organism
Gonads (testes and ovaries)	Organs for sexual reproduction; *Hydra* are dioecious, meaning that an organism has either testes or ovaries, but not both (i.e., male or female)
Basal disc	Specialized region for attachment to the substrate
Epidermis	Outer tissue layer; specialized for protection
Gastrodermis	Inner tissue layer; specialized for digestion
Mesoglea	Inert, acellular, jelly-like substance that aids in supporting the body

Check Your Progress

1. In which tissue layer are the cnidocytes located? Why?

2. Does *Hydra* reproduce sexually? Asexually? Give evidence to support your answers.

3. What type of digestion is *Hydra* capable of: intracellular, extracellular or both?

COLONIAL HYDROZOANS

Now turn your attention to a more typical member of the class Hydrozoa. Members of the genus *Obelia* are colonial hydrozoans that are connected by branches of a common gastrovascular cavity (**coenosarc**), making them all part of a larger functioning body. Due to this cooperative venture, certain polyps develop into highly specialized feeding polyps, while others lose the ability to feed altogether in exchange for an enhanced ability to reproduce. Despite its appearance, *Obelia* is not a true colony since the numerous polyps all form from the repeated budding of a single individual.

1. Obtain a w.m. slide of a colony of *Obelia* polyps. Examine it using low power and identify the following structures depicted in Figure 6.3 and defined in Table 6.2: **hydranth** (feeding polyp), **tentacles, hypostome, gonangium** (reproductive polyp), **medusa buds, perisarc** and **coenosarc.**

2. Now examine a w.m. slide of an *Obelia* medusa using medium power. Locate the following structures which are defined in Table 6.2: **tentacles, manubrium, mouth** and **gonads.** The medusa stage of most hydrozoans is typically a very short-lived stage devoted primarily to reproduction. Male and female medusae release haploid sperm and eggs, which fuse to form a diploid zygote. This zygote develops into a swimming **planula larva** which settles to the bottom of the ocean floor where it develops into a new polyp. Through this mechanism, *Obelia* has an alternation of generations from polyp to medusa and back to polyp again in a continuous cycle (Fig. 6.3).

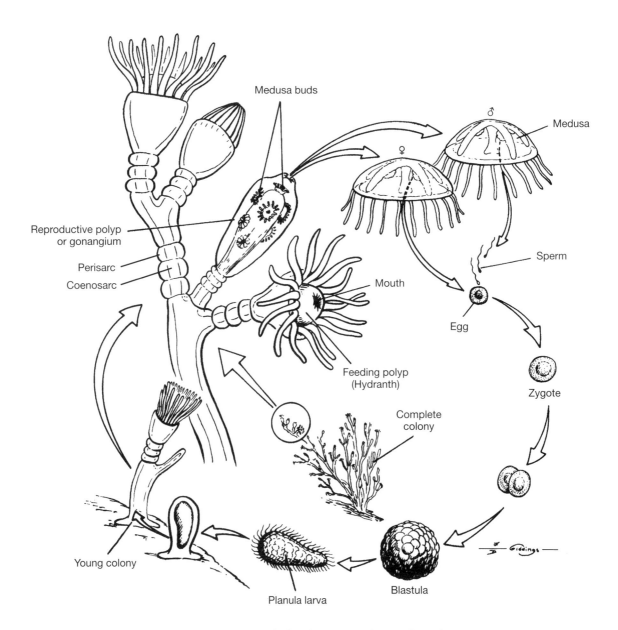

FIGURE 6.3 ▪ Lifecycle of *Obelia* showing polyp and medusa stages.

TABLE 6.2 ▪ Anatomy of *Obelia*

Body Form	Structure	Function
Polyp	Hydranth (feeding polyp)	Polyp specialized for food acquisition
	Tentacles	Defense and prey capture
	Hypostome	Elevated mound of tissue which expands or contracts to regulate size of mouth opening
	Gonangium (reproductive polyp)	Polyp specialized for reproduction
	Medusa buds	Product of asexual reproduction; medusae will be released from the gonangium when mature and will produce either sperm or eggs which fuse with the respective gamete forming a zygote which will develop into a new polyp
	Coenosarc	Common chamber within which extracellular digestion occurs; nutrients are distributed throughout organism
	Perisarc	Translucent outer covering of organism; serves protective function
Medusa	Tentacles	Defense and prey capture
	Manubrium	Stalk of fleshy tissue which supports the mouth
	Mouth	Ingestion of food and egestion of indigestible particles
	Gonads	Organs for sexual reproduction; either testes or ovaries

Check Your Progress

1. List several ways in which *Obelia* differs from *Hydra*.

Scyphozoan Anatomy

Materials needed:

▪ preserved scyphozoans
▪ dissecting microscope

Members of the class Scyphozoa are called jellyfish because of their thick layer of gelatinous mesoglea. In this class, the medusa stage dominates the life-cycle, with the polyp stage being relegated to an inconspicuous, short-lived larval form which quickly matures into a polyp that buds off young medusae. *Aurelia*, the moon jellyfish, is a common, widely-distributed genus within this class that typifies the anatomy of scyphozoans. In the open oceans where they thrive, they may reach sizes of up to 1 foot in diameter. You will not need to dissect these specimens, since their transparent body readily shows most of their anatomy.

1. Obtain a preserved specimen of *Aurelia* (or some other scyphozoan medusa).

2. Carefully examine it and identify the following structures using Figure 6.4 (anatomy defined in Table 6.3): **mouth, oral arms, marginal tentacles, gonad, gastric pouch, radial canals, circular canal.**

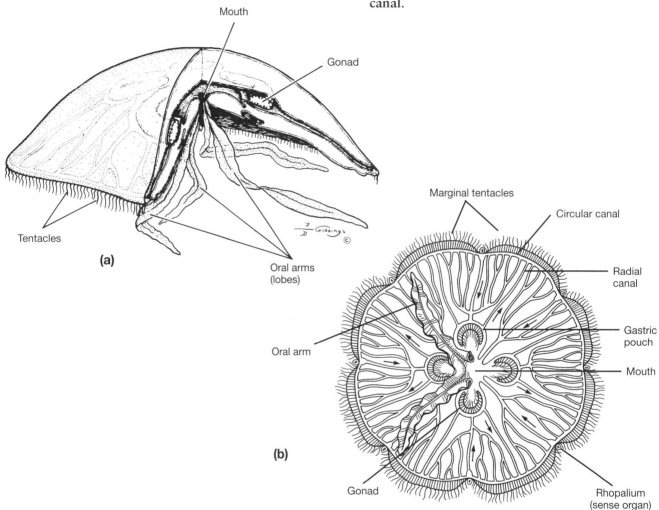

FIGURE 6.4 ▪ Lateral view (a) and ventral (oral) (b) view of a scyphozoan jellyfish.

TABLE 6.3 ▪ Anatomy of *Aurelia*

Structure	Function
Marginal tentacles	Provide sensory information and used in defense and locomotion
Mouth	Ingestion of food and elimination of indigestible particles (egestion)
Oral arms	Defense and prey capture
Gonads (testes and ovaries)	Organs for sexual reproduction; gametes are released into the gastric pouches and exit the body through the mouth
Gastric pouch	One of four divisions of the gastrovascular cavity for digestion of food
Radial canals	Extensions of the gastric pouches that radiate outward from the pouches and distribute nutrients throughout body
Circular canal	Circular extension of the gastric pouches that distributes nutrients to outer rim of jellyfish

Check Your Progress

1. Compare and contrast the scyphozoan medusa with the medusa bud of *Obelia*.

Anthozoan Anatomy

> *Materials needed:*
> - preserved sea anemones
> - dissecting microscope

Anthozoans are represented by sessile organisms in the polyp stage existing as solitary individuals (sea anemones) or as true colonies of dozens to thousands of individuals (corals) (Fig. 6.5). There is no free-swimming medusa stage in their lifecycle. Of the cnidarians, the anthozoans are the most specialized and their anatomy reflects this highly specialized life. You will examine a member of the genus *Metridium*, a common North Atlantic sea anemone. Most of the internal anatomy can be viewed in specimens that have been bisected longitudinally, so further dissection is probably not necessary.

1. Obtain a preserved specimen of *Metridium* that has been bisected longitudinally.

2. Examine this specimen and locate the following structures depicted in Figure 6.6 (anatomy defined in Table 6.4): **tentacles, oral disc, mouth, ostium, pharynx, retractor muscles, gonad, gastrovascular cavity** (coelenteron), **primary septum, secondary septum, pedal disc, acontia.**

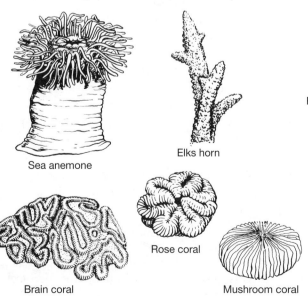

FIGURE 6.5 ▪ Representative anthozoans.

Sea anemone

Elks horn

Brain coral

Rose coral

Mushroom coral

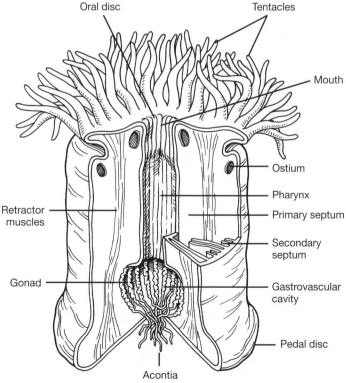

Oral disc

Tentacles

Mouth

Ostium

Pharynx

Primary septum

Secondary septum

Retractor muscles

Gonad

Gastrovascular cavity

Pedal disc

Acontia

FIGURE 6.6 ▪ Anatomy of dissected sea anemone (*Metridium*).

TABLE 6.4 ▪ Anatomy of *Metridium*

Structure	Function
Tentacles	Defense and prey capture
Mouth	Ingestion of food and elimination of indigestible particles (egestion)
Oral disc	Raised portion of the mouth (equivalent to the hypostome of hydrozoans)
Gonads (testes and ovaries)	Organs for sexual reproduction; gametes are released into the gastric pouches and exit the body through the mouth
Ostium	Pore that allows circulation of fluids between adjacent body sections
Pharynx	Muscularized portion of the gastrovascular cavity for pulling prey inward and expelling indigestible particles
Retractor muscles	Expand and contract the body
Gastrovascular cavity (coelenteron)	Specialized chamber for extracellular digestion of prey
Primary septum	One of several thin vertical walls that divide body into sections and provide support
Secondary septum	One of several thin vertical walls that further divide body and provide support
Pedal disc	Tough, fleshy base that attaches organism to rocks or sandy ocean floor
Acontia	Contain cnidocytes and aid in subduing live prey taken into the gastrovascular cavity

EXERCISE 6–4

Investigation of Cnidarian Feeding Behavior

Materials needed:

- live *Hydra littoralis*
- live blackworms (*Lumbriculus variegatus*)
- dropper bottle of tyrosine solution (0.9 g/L)
- dropper bottle of glutathione solution (1.53 g/L)
- dropper bottle of distilled water
- 5 Syracuse dishes
- 50 mL beaker
- disposable pipettes
- stopwatches
- dissecting microscope

Behavior is simply an animal's response to stimuli in its environment, and the basic physiological unit of behavior is the motor system. At its simplest level, the motor system consists of sensory structures, neurons and muscles (or glands) whose actions are coordinated to bring about certain behaviors.

When zoologists study behaviors, they distinguish between two basic causes of a behavior. The **proximate cause** of a behavior is the series of immediate physiological events that led up to the particular behavior. The **ultimate cause** of a behavior is the resultant evolutionary advantages that have promoted that behavior to remain in the animal's repertoire of possible responses. Think of proximate causes as explaining *how* the behavior occurs and ultimate causes as explaining *why* the behavior exists. A quick example should remove any confusion. Some species of night-flying moths have ears located behind their wings that can only detect sounds in the ultrasonic range (20 KHz – 40 KHz). This is well outside the range of sounds that the moths themselves can produce, making these moths deaf to their own sounds! It is, however, within the exact range that night-flying bats emit their sonar pulses. Thus when these moths detect pulses of sound in the ultrasonic range, they instantaneously initiate a complex series of aerial avoidance maneuvers involving flips, spins and dives that allows them to avoid becoming a late night snack for the bat.

What is the proximate cause of this behavior? Sensory receptors in the moth's ears are stimulated by the sound vibrations, causing action potentials in the moth's auditory neurons to be generated and propagated through the nervous system to motor neurons in the moth's flight muscles, bringing about the subsequent array of aerial gymnastics. Does the moth have to think about what it's doing? Of course not; this is a "hard-wired" behavior. It occurs whether there's actually a bat making the sound or a sophisticated machine in the laboratory producing the high-frequency pulses. What then, is the ultimate cause of this behavior? Since night-flying bats do produce sounds in these frequencies while hunting, moths that (1) had some capacity for detecting these sounds and (2) had reflexive avoidance behaviors in response to these sounds enjoyed greater survival than their less-fortunate relatives and consequently left more offspring of their own that had the genetic predisposition for the right ears and the right moves.

The cnidarian neuromuscular system is far simpler than that of a moth, or a bat for that matter, but it represents an ideal model for analyzing the general principles of simple behaviors. When an object brushes against one of its tentacles, *Hydra* reacts by "stinging" the object with its nematocysts. If that object is a small, consumable food item, the nematocysts will pierce through the body of the prey and cause the release of fluids from within the prey. *Hydra* have the capacity to sense the presence of certain chemicals in their environment through receptor proteins on the surface of specialized sensory neurons in their epidermis.

The chemicals glutathione and tyrosine are two amino acids that are present in most living creatures, including small aquatic worms and arthropods. Are the receptors of *Hydra* sensitive enough to distinguish between these two very small molecules? (Glutathione is a tripeptide: Glu-Cys-Gly, while tyrosine is a single amino acid.) Furthermore, are amino acids the proximate cue that *Hydra* use to differentiate between small, consumable prey and nonconsumable "trash" that just happened to bump into their tentacles? The following exercise uses the scientific method to answer these questions and determine if the amino acids glutathione and tyrosine are the proximate triggers of *Hydra's* feeding response.

Before proceeding, your instructor should discuss the background regarding this experiment and you should devise a good scientific question and testable hypothesis for the experiment. You will set up 5 dishes, each containing 1–2 *Hydra*. Each dish will receive one of the following substances: (1) distilled water, (2) sections of blackworm, (3) glutathione, (4) tyrosine or (5) glutathione + tyrosine. You will categorize any behaviors seen and record the degree and duration of each response. Use the following list of behaviors that you may observe during a normal feeding event to categorize the behaviors seen during the experiment. To quantify these behaviors you should record (1) the intensity (on a scale of 1–10), (2) the frequency and (3) the duration of each reaction.

Typical feeding behaviors include:
- tentacles waving randomly
- tentacles drawn toward mouth
- hypostome expansion (mouth open)
- hypostome contraction
- body contraction
- body elongation

Using the Scientific Method

1. What is the general question being addressed by this experiment?

2. Record the hypothesis you wish to test.

3. Identify the independent variable in this experiment.

4. Identify the dependent variable(s) in this experiment.

5. What were the levels of treatment (i.e., levels of the independent variable)?

6. What served as the control treatments (Hint: there was a positive control and a negative control)?

7. Identify several controlled variables.

Experimental Protocol

1. Work in groups of 4 people for this experiment. Obtain 5–6 living *Hydra* from your instructor for your group.

2. Place your *Hydra* individually in a small amount of pond water in 5 Syracuse dishes.

3. Adjust the mirror and lighting on your dissecting scopes to achieve optimum illumination of the specimen. These animals are likely to contract into a ball shape as a result of being transferred to a new environment. Therefore, you should transfer the animals to your experimental setup (i.e., dish and microscope) and then turn off the microscope light. Leave the *Hydra* undisturbed for 4–5 minutes.

4. While the *Hydra* are "resting," use this time to prepare small sections of the blackworms for feeding. Since adult blackworms are a bit too large for our *Hydra*, you will need to slice them into smaller pieces and hand feed them to the *Hydra* using extremely fine-tipped forceps.

5. Place a blackworm on wet filter paper on the stage of a dissecting microscope and slice it into sections that are 3–4 segments long (≈1 mm long) using a sharp razor blade. The wet paper inhibits its undulating movements, making slicing easier.

6. After 4–5 minutes, return to the dissecting microscope and observe one of the *Hydra* with subdued light. Can you see the tentacles waving?

7. One *Hydra* per group will serve as the control treatment for the experiment. Using the dropper bottle place 3 drops of distilled water into the dish near the animal. What is the response? How long does the response to the disturbance last?

8. Record your observations in Table 6.5.

TABLE 6.5 ▪ Effects of Distilled Water, Amino Acids, and Live Prey on Feeding Responses in *Hydra*

For each treatment, you should record (1) the intensity (on a scale of 1–10), (2) the frequency, and (3) the duration of each response.

Potential Behaviors	Treatments				
	Distilled Water	Blackworm Fragments	Glutathione	Tyrosine	Glutathione + Tyrosine
Tentacles waved randomly					
Tentacles drawn toward or across mouth					
Hypostome expansion (mouth open)					
Hypostome contraction (mouth closed)					
Body elongation					
Body contraction					

9. Repeat the stimulus and note how long it takes the *Hydra* to recover.

10. Use fine forceps to carefully position a small piece of the blackworm next to one of the outstretched tentacles of the *Hydra* and touch the worm segment to the tentacle. What is the immediate reaction? Does it differ from the reaction to distilled water? If the distilled water control animal has recovered sufficiently, it can also be fed.

11. Watch very carefully to observe the natural feeding behaviors of *Hydra*. These behaviors will serve as the "standard" to which you will compare the *Hydra's* response to the different chemicals.

12. Record your observations in detail in Table 6.5.

13. To one of the other *Hydra* in a separate dish, add 3 drops of the amino acid glutathione. Observe the reaction carefully.

14. Record your observations in Table 6.5.

15. To another *Hydra* in another dish, add 3 drops of the amino acid tyrosine. Observe the reaction carefully.

16. Record your observations in Table 6.5.

17. To the remaining *Hydra* in the last dish, add 3 drops of glutathione, *wait 4–5 minutes*, then add 3 drops of tyrosine. Observe the reaction carefully.

18. Record your observations in Table 6.5.

Check Your Progress

1. Does *Hydra* respond to chemical cues in its environment?

2. Is the feeding behavior an all-or-nothing response or can you distinguish discrete phases of the behavior?

3. What conclusions can you draw about the kinds of receptors involved in the induction of feeding behavior and their distribution in the body?

4. Are amino acids the proximate cause of the feeding response seen in *Hydra*?

5. Describe the normal feeding behavior of *Hydra* in terms of physical and chemical stimulation.

Questions For Review

1. What specialized cells in *Hydra* aid in capturing and subduing prey?

2. What structures determine whether a polyp of *Obelia* is a hydranth (feeding polyp) or a gonangium (reproductive polyp)?

3. What modification do sea anemones possess to allow food to be distributed among the partitioned regions of their bodies?

4. Match the class with the correct organism or characteristic.

   _____ a. *Hydra*   1. Class Hydrozoa

   _____ b. *Obelia*   2. Class Scyphozoa

   _____ c. sea anemone   3. Class Anthozoa

   _____ d. jellyfish

   _____ e. no medusa form; only polyp

   _____ f. medusa form dominant; polyp greatly reduced

   _____ g. both polyp and medusa forms

5. Explain the difference between the proximate cause of a behavior and the ultimate cause.

6. As you walk around your backyard on a damp night with a flashlight pointed at the ground, you notice all of the earthworms quickly recoiling into their burrows as you walk near them. Speculate about the proximate and ultimate causes of this stereotypical behavior.

Platyhelminthes

After completing the exercises in this chapter, you should:

1. Understand the three types of body plans that characterize triploblastic animals.
2. Recognize the basic anatomy of common members of this phylum.
3. Be able to distinguish between the three major classes of flatworms.
4. Understand the evolutionary modifications for parasitic or free-living lifestyles that members of each class possess.
5. Understand the basic behavioral patterns of planaria and the degree to which these behaviors can be modified through learning.
6. Understand the concept of cephalization and its evolutionary significance.
7. Understand the concept of regeneration and the degree of flexibility present in flatworms for regenerating lost body parts.
8. Be able to define all boldface terms.

ANIMAL BODY PLANS

Animals, such as flatworms, containing three layers of tissue are known as **triploblastic**; remember cnidarians are diploblastic and sponges have no true tissues. In fact most animals with which we are familiar are triploblastic. Within the triploblastic animal species different developmental strategies have evolved for packaging the tissues and compartmentalizing the body space. As a result, three different body plans have arisen: acoelomate, pseudocoelomate and eucoelomate (Fig. 7.1).

Body Plans of Triploblastic Animals

Acoelomate—animals whose central space is filled with tissue (mesoderm). No true body cavity exists. Example: flatworms.

Pseudocoelomate—animals with a central body cavity that lies between gastrodermis and mesoderm. Example: roundworms (nematodes).

Eucoelomate—animals with a central body cavity that lies within mesoderm. Examples: earthworms, molluscs, insects, chordates.

Remember that the gut (or digestive cavity) does not count as a body cavity when determining the type of coelom that an animal possesses. All triploblastic animals possess a digestive cavity. Some, such as pseudocoelomate and eucoelomate animals, possess an additional space within the body—the coelomic space. Others, the acoelomates, lack this additional body cavity.

There are an estimated 15,000 species of flatworms. They range in size from a few millimeters in length to over 20 meters long! Despite this wide range in length, flatworms are never more than a few millimeters thick. Their anatomy differs substantially from the cnidarians you observed in the previous chapter. In addition to the **epidermis** and **gastrodermis**, flatworms possess a third tissue layer sandwiched between the two called **mesoderm**. During embryonic development this third tissue layer differentiates into muscles (among other things), making flatworms one of the first highly motile groups of animals. Flatworms have **bilateral symmetry**, and free-living species typically possess a concentration of nervous tissue and sensory structures at the cranial end of the body—a condition known as

DATE: **DIAGRAM OF PLANARIA**   **DESCRIPTION OF PROGRESS:**

Questions For Review

1. Contrast the nervous system seen in planaria (*Dugesia*) with that seen in *Hydra*.

2. Distinguish between the processes of egestion (or defecation) and excretion using the flatworm as a model for both processes.

3. Define cephalization.

4. What is the evolutionary advantage for bilaterally-symmetrical, motile animals such as flatworms to have a concentration of nervous tissue and sensory organs located at their cranial end?

5. The cuticle of flukes is highly resistant to enzyme action. Why is this an important evolutionary modification for endoparasites?

6. Why does the tapeworm lack a mouth, well-developed sensory structures and a digestive system?

7. Based on your observations of planaria behavior, predict how a planaria may react to each of the following stimuli and indicate which sensory structure would be responsible for the response.

    a. A bright, sunny day.

    b. A dead earthworm at the bottom of the lake.

    c. A small minnow nibbling at the flatworm's tail.

Nematoda

After completing the exercises in this chapter, you should be able to:

1. Discuss the major characteristics of nematodes.

2. Recognize the basic external and internal anatomy of a common nematode.

3. Characterize locomotion in roundworms and describe the effects of temperature on the metabolic rate of ectothermic organisms.

4. Define all boldface terms.

The phylum Nematoda is composed of a mixture of free-living and parasitic members. Nearly 20,000 species have been described in this vast phylum, but biologists estimate that there may be 4–50 times that number in existence. In fact nematodes are so numerous that if you were to remove everything else from the environment, the remaining nematodes would form a ghostly skeleton outlining the entire terrestrial biosphere! Nematodes possess two major morphological advances over the flatworms you studied in the previous chapter. They have a **pseudocoelom**, a body cavity that lies between a layer of mesoderm and a layer of gastrodermis, and a **complete digestive tract** containing both a mouth and an anus. The evolutionary advantages of each of these characteristics are discussed later. Whereas flatworms show tremendous diversity in body form, nematodes tend to display more similarities than differences among the numerous species. They all possess unsegmented, tapered, tubular bodies devoid of appendages and covered by a thin **cuticle** secreted by the epidermis. The cuticle is permeable only to water, gases and some ions and thus serves as a protective coating, especially in parasitic forms. Another difference between nematodes and flatworms is that in nematodes the sexes are usually separate (**dioecious**) and females are generally larger than males.

A unique characteristic of this phylum is that each animal has a finite number of cells in its body and this number is species specific (meaning that every member of the same species has the same number of cells, but different species may have different cell numbers). Cell division stops fairly early during embryonic development in nematodes and future growth during later developmental stages is achieved through cell growth (rather than through cell division). One common free-living nematode, *Caenorhabditis elegans*, is transparent throughout development, making it possible for biologists to trace the lineage of every cell from the zygote to the adult worm. Biologists have constructed a fate map of the lineage of all 959 cells in this roundworm and have recently sequenced its entire genome, making it the first multicellular organism to have its entire genetic sequence decoded.

Although nematodes are usually only a few millimeters thick, they may range in length from a few millimeters to several meters. One parasitic species that inhabits the placenta of female sperm whales may attain a length of 9 meters! The infectious potential of parasitic nematodes is staggering. For instance, in Africa alone, 40 million people are presently infected with the larva of *Onchocerca spp.*, a nematode which migrates to the victims' eyes causing blindness. *Onchocerca* infection is currently one of the major causes of blindness worldwide. The disease commonly referred to as elephantiasis results from a blockage of the lymphatic system brought about by one of several nematode species. This affliction causes substantial buildup of fluid and subsequent dense growth of connective tissue, grossly deforming various body regions. This condition is one of the world's fastest spreading diseases and presently afflicts nearly 100 million people. Pinworms affect some 500 million people, particularly children, in temperate regions of the globe.

Not only are their sheer numbers bewildering, but their ability to reproduce in large numbers is equally impressive. A single female *Ascaris* releases some 200,000 fertilized eggs per day (that's 73 million per year!). Females of the guinea worm, *Dracunculus medinensis*, live just beneath the skin in their human hosts. When the skin comes into contact with water (as when the host bathes), the nematode protrudes its caudal end through the sore on the host's skin and can eject up to 1.5 million offspring into the water during the course of a day. These offspring cannot directly reinfect humans, but instead carry out the lifecycle in a species of microscopic aquatic crustacean. Humans are infected when they drink the water containing these tiny arthropods. The cure for this infection is simple but tedious. An incision is first made in the skin of the host near the sore and the one-meter worm is slowly rolled out on a match stick a few centimeters daily until the entire worm has been removed.

EXERCISE 8–1   **Photographic Atlas Reference Pages 60–61**

Nematode Anatomy

> *Materials needed:*
> - preserved *Ascaris lumbricoides*
> - dissection tools
> - dissecting scope
> - dissecting pan
> - slide of *Ascaris* male and female c.s.
> - compound microscope

1. Obtain a preserved specimen of the roundworm *Ascaris lumbricoides*. This is a large, parasitic roundworm that infects humans and pigs. *Be very careful when handling preserved nematodes.* Nematode eggs are extremely resilient and may remain viable even after the females have been preserved. Keep your hands away from your eyes, nose and mouth to avoid possible contamination.

2. First determine the sex of your nematode. Males are typically smaller, have a curved caudal end and have two small, spiny projections called **spicules** on the ventral surface near the anus. These spicules are used during copulation.

3. After you have identified the sex of your specimen, differentiate between the cranial and caudal ends. This is more apparent on the males due to their curved tails. The cranial end of both sexes is generally more pointed than the more blunt caudal end. The anus is located on the ventral surface rather than at the terminal portion of the tail—another useful feature for distinguishing which end is which.

4. Use a dissecting scope to examine the cranial end of your specimen. Notice the **triradiate mouthparts** (lips) surrounding the mouth. This is a distinguishing feature of the phylum.

5. Use a dissecting needle to gently scrape away a piece of the thin cuticle from the body. Notice that it is composed of a transparent, proteinaceous substance. In parasitic species such as *Ascaris*, this cuticle protects the animals from the digestive enzymes of the host.

6. Submerge your specimen in water in your dissecting pan and use a sharp dissecting needle to make an incision through the body wall of the nematode near the mouth. Keeping the roundworm immersed in water, draw the needle along the length of the roundworm, extending the incision the entire length of the body.

7. Use several pins to carefully open your specimen and secure it to the pan. HINT: Pin your specimen near one edge of the dissecting pan so that you can easily view it under the dissecting scope.

8. Use Figure 8.1 and Table 8.1 to identify the internal anatomy of *Ascaris*.

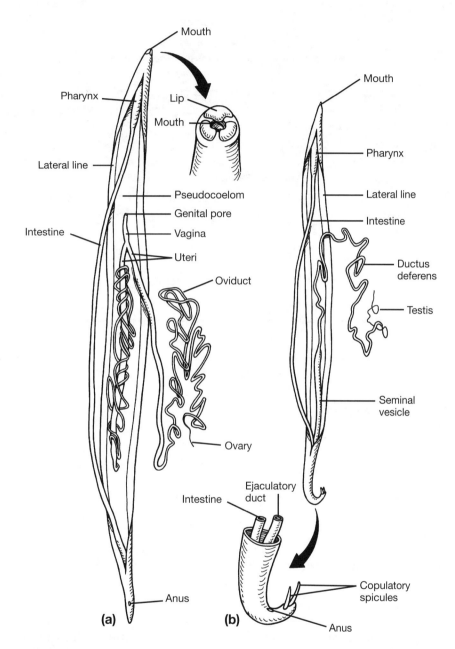

FIGURE 8.1 ▪ Anatomy of a dissected nematode (*Ascaris lumbricodes*). Female (a) and male (b) depicted separately.

TABLE 8.1 ▪ Anatomy of a Roundworm (*Ascaris*)

Structure	Function
Mouth	Ingestion of food
Pharynx	Muscularized region of digestive tract that "pumps" food through the mouth and into the intestine
Intestine	Ribbon-like digestive tract where absorption of nutrients occurs
Anus	Elimination of indigestible wastes (egestion)
Lateral lines	Longitudinal canals that function as the excretory system of the roundworm, releasing nitrogenous wastes in the form of ammonia and urea
Pseudocoelom	Body cavity lined on the inside by a layer of gastrodermis and on the outside by a layer of mesoderm
Testis (male)	Produces sperm
Ductus deferens (male)	Stores mature sperm and transports them to seminal vesicle
Seminal vesicle (male)	Enlarged tube representing terminal portion of male reproductive tract which transports mature sperm out of the nematode
Genital pore (female)	Point of entry for sperm and opening through which fertilized eggs are released from the body
Vagina (female)	Terminal portion of female reproductive tract which receives sperm from males and directs eggs through genital pore
Branched uterus (female)	Site where developing eggs mature before being released
Oviduct (female)	Repository for eggs produced in ovary until fertilization
Ovary (female)	Produces eggs

Check Your Progress

1. Is your nematode male or female? How do you know?

2. Is the thin, transparent cuticle evident on your specimen? (Look along the edge of the incision in the body wall.)

3. Is the intestine surrounded by an outer layer of muscle? (HINT: Think of the type of body cavity this animal possesses.)

4. Do nematodes fit the pattern typical of parasites by having a large portion of their body cavity devoted toward reproductive structures?

5. Through which structure are male gametes released?

6. Through which structure are females' fertilized eggs released?

EXERCISE 8–2

Locomotion in the Vinegar Eel

> *Materials needed:*
> - live vinegar eels (*Anguillula aceti*)
> - test tubes w/stoppers
> - water baths at 10°C, 15°C, 20°C and 25°C
> - dissecting microscope

1. Place a few vinegar eels into a clear test tube and insert the stopper.

2. Using the dissecting scope, observe their pattern of locomotion. How would you describe this pattern?

3. Most nematodes display a characteristic undulating, whip-like pattern of movement. By focusing your attention on one small portion of the roundworm's body (e.g., the head or tail), it is possible to count the number of undulations over a given time frame. Practice counting the number of undulations in your specimen over 30-second time intervals. Do this several times and record the number of undulations for each time interval below.

Time Interval #1: _____ (no. of undulations)

Time Interval #2: _____ (no. of undulations)

Time Interval #3: _____ (no. of undulations)

Time Interval #4: _____ (no. of undulations)

Check Your Progress

1. Are the number of undulations fairly consistent between trials?

2. What is the average undulation rate of vinegar eels under normal conditions? (Remember to express the rate as a function of time, i.e., # undulations/minute.)

3. Based on your knowledge of nematode anatomy, what kinds of muscles are involved in producing these movements?

Now that you have a method for reliably quantifying the locomotive abilities of this animal, you will investigate the effects of a changing environmental parameter on locomotion in vinegar eels. You will subject vinegar eels to different environmental temperatures and record their undulation rates to see if temperature affects locomotion. Four test tubes each containing 5 vinegar eels will be placed in water baths set at (or near) the following temperatures: 10°C, 15°C, 20°C and 25°C. You will record the undulation rate of each vinegar eel for 30 seconds (for a total of 20 recordings—5 roundworms at each of 4 different temperatures). Like all invertebrates, roundworms are **ectothermic**, meaning they do not generate body heat internally. The label "cold-blooded" is not an appropriate term for describing ectothermic organisms, for many ectotherms actually sustain body temperatures well above that of most birds and mammals (the traditional "warm-blooded" creatures). Ectothermic simply means that the source of body heat for an organism is outside the body. Birds and mammals are endothermic, meaning that their body heat is generated from within. It stands to reason then that an ectothermic animal such as a roundworm may be profoundly affected by changes in its surrounding environmental temperature. Such changes may affect its behavior, metabolic rate, growth rate and even its basic biochemical pathways. These features make ectothermic animals excellent models for elucidating the complex biological relationships between an animal and its thermal environment.

Using the Scientific Method

1. What is the general question being addressed by this experiment?

2. Record the hypothesis you wish to test.

3. Identify the independent variable in this experiment.

4. Identify the dependent variable in this experiment.

5. What are the levels of treatment (i.e., levels of the independent variable)?

6. What served as the control treatment?

7. Identify several controlled variables.

Experimental Protocol

1. Work in groups of 4 people for this experiment. Obtain 20 test tubes each containing a vinegar eel.

2. Label 5 test tubes 10°C, 5 tubes 15°C, 5 tubes 20°C and 5 tubes 25°C.

3. Place an appropriately labeled tube in each of the four water baths and let the vinegar eels habituate to their new thermal environment for 3–4 minutes.

4. Have each person in your group record the undulation rate of one vinegar eel for 30 seconds and record the data in Table 8.2.

5. Repeat steps #3 and #4 four more times, *using a new vinegar eel in a new test tube each time* for each of the 4 temperature settings (for a total of 20 recordings—5 roundworms at each of the 4 different temperatures).

6. Record your data in Table 8.2.

TABLE 8.2 ▪ Locomotion in the Vinegar Eel (*Anguillula aceti*)

Record the number of undulations per 30-second interval for 5 roundworms in each of the 4 temperature settings.

	Temperature			
Specimen #	10°C	15°C	20°C	25°C
1				
2				
3				
4				
5				
Average undulation rate (# undulations/30 sec.)				

Check Your Progress

1. Convert your results to the average number of undulations per minute for each of the four temperatures.

   10°C =

   15°C =

   20°C =

   25°C =

2. Does environmental temperature affect locomotion in ectothermic animals such as vinegar eels?

3. Discuss the effects on locomotion that variation in the thermal environment of vinegar eels produces.

4. Based on your data, should you retain or reject your hypothesis?

Questions For Review

1. List two distinct advantages that roundworms possess over flatworms.

2. What are the advantages of a digestive tract having a separate entrance and exit?

3. For a parasitic organism like *Ascaris*, what would be the selective advantage of a cuticle?

4. For a free-living, terrestrial nematode, what would be the selective advantage of a cuticle?

5. In many parasitic species, the sex ratio of males to females is skewed (i.e., not 1:1). Which sex do you suppose would tend to be more numerous? Why?

6. Explain the difference between a pseudocoelom and a true coelom.

7. Define ectothermic.

8. The hookworm, *Necator americanus*, which infects some 900 million people worldwide may ingest more than 0.5 mL of human host blood daily. Given that infections may number over 1000 individuals, calculate the total volume of host blood that may be lost per day from a severe nematode infection.

9. Given that the total blood volume of the average adult human is 5 L, calculate the percentage of total blood volume lost daily in the example above.

Mollusca

9

After completing the exercises in this chapter, you should be able to:

1. Discuss the major characteristics of molluscs.
2. Identify the major anatomical features of a freshwater mussel.
3. Explain the pattern of water flow through a freshwater mussel and its significance.
4. Identify the major anatomical features of a squid.
5. Compare and contrast the evolutionary modifications of filter-feeding molluscs with predatory molluscs.
6. Discuss the concept of adaptive radiation and explain how molluscs illustrate this phenomenon.
7. Define all boldface terms.

The phylum Mollusca is a large and remarkably diverse collection of animals that includes clams, mussels, scallops, oysters, snails, slugs, octopuses, squids and the chambered nautilus. There are nearly 50,000 described species of molluscs and potentially 100,000–150,000 species in existence today. Despite the diversity that exists within this phylum, the basic body plan of all molluscs is relatively similar. They are bilaterally symmetrical, unsegmented organisms that have a true coelom and well-developed organ systems. Molluscs possess four major morphological features that distinguish them from other invertebrates: (1) a protective **shell** (reduced in some species), (2) a **mantle**, (3) a **visceral mass** which houses the major internal organs and (4) a **foot** for locomotion. These basic body parts are modified in different ways in molluscs and these morphological variations clearly illustrate how diversity can be achieved by alterations to a relatively simple body plan. For example, the degree of cephalization varies enormously within this phylum, ranging from a total lack of cephalization in clams and other bivalves to the well-developed brain and image-forming eyes of squids and octopuses.

Because of the extreme diversity of molluscan body forms it may at first glance seem that these animals do not belong in the same phylum. Such a vast range of morphological types within a group of organisms sharing a common lineage is an example of **adaptive radiation**—the evolution of numerous species from a common ancestor following migration into a new environment. Adaptive radiation is often accelerated when the new environment has very few existing competitors and the introduced organisms are able to quickly disperse into different areas. Over many generations, these organisms become highly specialized to their particular surroundings and divergent forms arise. Molluscs branched off the main animal line over 500 million years ago during a period when very few other large organisms occupied the oceans. Competition for resources was relatively low and there were numerous ecological niches that were either unfilled or easily conquered by these newcomers. Keep in mind that at this time there were also no plants or animals living on land; most life was confined to water. In fact, plant and animal life on land would not arrive for another 50 million years! Thus today's living molluscs represent descendants of a very ancient group of organisms—a group that has experienced hundreds of millions of years of gradual evolutionary change. In the ensuing time frame, molluscs have become adapted for virtually every type of freshwater and marine habitat as well as numerous terrestrial habitats.

In the next two exercises you will examine the external and internal anatomy of two highly divergent molluscs: a freshwater mussel (or clam) and a squid. As you familiarize yourself with the anatomical structures of these organisms keep in mind how each animal's anatomy has been modified to allow it to be maximally adapted to its particular lifestyle.

Clams and mussels are members of the class **Bivalvia** and are generally stationary filter-feeders that burrow beneath the sand and extend their siphons through the sand to filter water into their bodies and extract oxygen and nutrients. Squid are members of the class **Cephalopoda** which also includes the octopus and nautilus. All cephalopods are highly-active, visually-oriented predators that hold their own in competition with some of the fiercest predators of the oceans.

EXERCISE 9–1   **Photographic Atlas Reference Pages 49–50**

Bivalve Anatomy

Materials needed:
• preserved freshwater mussel
• dissection tools
• dissecting pan

1. Obtain a preserved specimen of the freshwater mussel *Anodonta* or other bivalve.
2. First use Figure 9.1a to identify the cranial, caudal, dorsal and ventral regions of the specimen. These areas will be important points of reference for identifying internal structures.

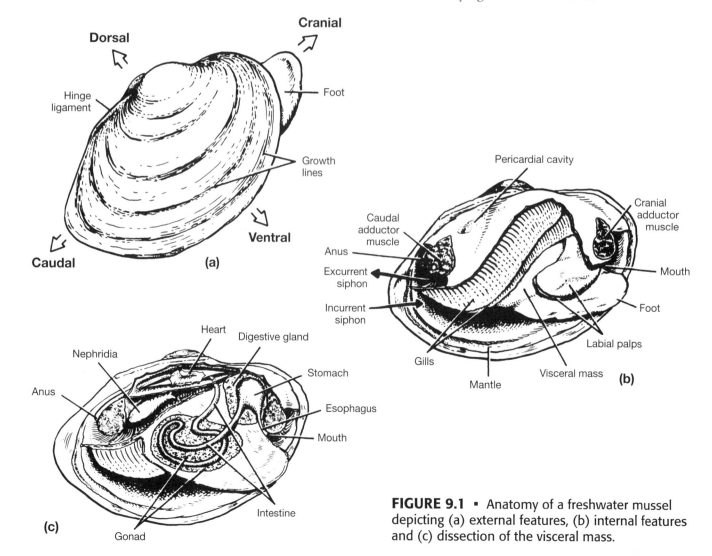

FIGURE 9.1 ▪ Anatomy of a freshwater mussel depicting (a) external features, (b) internal features and (c) dissection of the visceral mass.

3. Your specimen should have a wooden peg in it to keep the valves partially separated. Carefully insert a scalpel in the space between the right valve and the mantle near the wooden peg and slice through the nearby adductor muscle.

4. Repeat this procedure with the adductor muscle at the other end. This process requires a bit of feel to locate the muscles, since they are not visible through the small opening in the shell. Be careful that you do not damage any of the other internal organs while you are cutting the adductor muscles.

5. To completely open your specimen, it may be necessary to use a dissecting needle to gently peel the thin, fleshy mantle away from the shell.

6. Once it is open, lay your mussel down on its left shell (as pictured in Fig. 9.1b) and identify the structures indicated in the diagram and defined in Table 9.1.

7. Several major internal organs are located within the visceral mass. To dissect this structure use your scalpel to make a *longitudinal incision* through the visceral mass, dividing it into two bilaterally symmetrical halves (in the same plane that the shell opens). Use Figure 9.1c to help you identify the internal organs of the visceral mass. You should be able to locate **gonad** tissue, **intestinal coils**, **digestive gland** tissue and the **stomach**.

TABLE 9.1 ▪ Anatomy of a Bivalve (Clam or Freshwater Mussel)

Structure	Function
Incurrent and excurrent siphons	Extendable, fleshy tubes that transport water into and out of the body
Gills	Used primarily for respiration and filter feeding; female freshwater mussels brood eggs in special gill pouches
Mantle	Thin membrane that secretes the shell
Shell	Hard outer covering that protects soft internal organs; composed of a mixture of calcium carbonate and protein
Foot	Muscularized region adjacent to visceral mass for burrowing and locomotion
Visceral mass	Pouch that contains several major internal organs
Adductor muscles	Large, tubular muscles located at the cranial and caudal ends of the animal; close shell and hold it tightly together
Labial palps	Fleshy folds of skin located near the mouth that collect food particles from the gills and transport them to the mouth
Mouth	Ingestion of food
Esophagus	Short tube through which food passes from mouth to stomach (rarely visible on dissection)
Stomach	Small chamber located within visceral mass for food storage
Digestive gland	Greenish, granular tissue that secretes digestive enzymes into stomach and intestine to assist in the breakdown of food
Intestine	Coiled digestive tract where absorption of nutrients occurs
Anus	Elimination of indigestible wastes (egestion)
Gonad	Produces gametes for reproduction
Heart	Muscularized portion of circulatory system that receives blood that collects in the open sinuses and pumps it through short arteries to neighboring tissues and organs
Nephridia	Excretory organs of bivalve which concentrate nitrogenous wastes and eliminate them from the body

FEEDING AND REPRODUCTION

Mussels and other bivalves depend on a constant flow of water through their bodies for oxygen and nutrient acquisition and for releasing gametes and wastes. As you examine your specimen, trace the flow of water through the body. Water enters through the more ventral incurrent siphon and immediately passes over the feathery gills. The gills extract oxygen and small, suspended food particles from the water while releasing carbon dioxide into the water. The gills have a thin coating of mucus which traps food particles and allows them to be passed by ciliary action toward the labial palps near the cranial end of the animal. Water circulates dorsally through the mantle cavity and makes a 180° turn, passing along the dorsal aspect of the mantle cavity. Here nitrogenous wastes are excreted by the nephridia into the water. As the water leaves the clam through the more dorsal excurrent siphon, it passes directly past the anus where wastes are eliminated from the digestive system and swiftly carried away from the animal.

In addition to respiration and food acquisition, the gills in female freshwater mussels play a role in reproduction (molluscs are **dioecious**, so the sexes are separate). Eggs are fertilized within special chambers in the female as sperm are brought in by water currents. Females then brood the fertilized eggs in special pouches in the gills until the eggs are ready to hatch. The eggs develop into tiny larvae called **glochidia** which attach to the gills of certain fish species and act as external parasites, stockpiling the necessary nutrients from the host fish to complete their embryonic development. Later, the parasitic larvae detach from the gills of the host fish and settle to the bottom to complete their transformation into free-living mussels. In marine bivalves, sperm and eggs are usually shed simultaneously and fertilization is external, resulting in the formation of free-swimming larvae that disperse and develop into mature adults.

CIRCULATION

Molluscs represent the first phylum you have studied whose members have a true circulatory system. Bivalves such as clams, mussels and oysters have an **open circulatory system**—a system in which the blood is not confined within a network of vessels. In mussels, blood from tissues and major organs flows to the gills where it is oxygenated and then collects in large, open sinuses that direct the flow of blood passively back to the heart. The oxygenated blood enters through openings in the heart called **ostia** and is pumped out of the heart through **arteries** to the mantle, foot and visceral mass. There are no veins present in this open system.

Check Your Progress

1. List several features of the freshwater mussel which enable it to thrive as a sedentary, aquatic organism.

2. Describe how the gills of a mussel are an example of a multifunctional organ.

EXERCISE 9–2   **Photographic Atlas Reference Page 51–52**

Cephalopod Anatomy

Materials needed:
- preserved squid
- dissection tools
- dissecting pan

As their name implies, members of the class Cephalopoda have a modified "head-foot" which bears an array of prehensile **tentacles** and arms at the cranial end of the body. The **visceral mass** is located toward the caudal end. Only the nautilus possesses an external shell; the shell is completely lacking in octopods and is reduced and internal in squids and cuttlefish. While squids and cuttlefish typically use their muscularized fins for leisurely locomotion, they also possess the ability to maneuver quickly by jet propulsion—rapidly expelling water from their mantle cavity through their tubular **siphon**. The evolutionary recruitment of the mantle cavity as the fluid reservoir for jet propulsion was incompatible with a hard external shell and explains why this feature is absent or highly reduced in most cephalopods. The consequence of this evolutionary trend was increased vulnerability to predation due to the lack of protection afforded by a shell. As a result, many octopods and squids developed the ability to rapidly change their skin color through the use of specialized epidermal cells allowing them to blend in perfectly with their surroundings. This, in turn, made them more effective predators. In addition, most cephalopods (except the nautilus) have an **ink sac** which can discharge a dark, cloudy liquid through the anus to confuse potential predators.

1. Obtain a preserved squid and position it in your dissecting pan so that the side with the siphon is toward you.

2. Examine the external anatomy of the squid and identify the following structures: **tentacles, arms, fins, siphon, mantle, eyes** and **collar** (Fig. 9.2a).

3. Using your scissors, make a shallow, longitudinal incision through the mantle starting at the collar and extending to the caudal end of the body tube (near the fins). You may need to use pins to keep the body tube open.

4. Use Figure 9.2b (if your squid is male) or Figure 9.2c (if your squid is female) and Table 9.2 to identify the following selected internal structures

of the squid and familiarize yourself with their functions: **gills, esophagus, stomach, pancreas, liver, anus, testis** (male only), **ovary** (female only), **ink sac, siphon retractor muscles, kidney, systemic heart, branchial hearts, caudal (posterior) vena cava,** and **cranial (anterior) vena cava.**

Cephalopods have the most efficient circulatory system of all molluscs, supporting the relatively high metabolic needs associated with swift-moving, active predators. They are unique among molluscs in having a **closed circulatory system** in which the blood is permanently contained within a network of **arteries** and **veins**. This advancement allows the blood to contain an unusually high concentration of oxygen-binding respiratory pigments. In addition to a single **systemic heart**, which receives oxygenated blood from the **gills** and sends it back to the tissues, squids have a pair of **branchial hearts** associated with the gills to increase blood pressure and thus blood supply to the capillaries of the gills.

Perhaps the single most remarkable adaptation that squids possess is their well-developed eye. The striking similarity between cephalopod eyes and the eyes of vertebrates (birds, mammals, fishes, etc.) is one of the most beautiful examples of **convergent evolution** among animals; each group has independently evolved acute, image-forming eyes that are amazingly similar in structure. The eye of the squid contains a lens, cornea, iris, ciliary muscles and a retina, just like the eyes of vertebrates. Due to their independent evolution, the cephalopod eye does differ in a few important ways from the vertebrate eye. In vertebrate eyes the lens is elastic and visual images are focused on the retina by altering the shape of the lens. In cephalopod eyes the lens is rigid and images are focused on the retina by altering the distance between the lens and retina (just like in a camera).

Another major difference between the two eyes is in the way light is received by the photoreceptors in the retina. In the vertebrate eye, the rods and cones point toward the back of the eye (away from the pupil) so light must pass through the photoreceptors and other associated nerve cells and bounce off the back of the retina (back toward the pupil) before it is detected by the photoreceptors! An unfortunate consequence of this design is that all of the neurons are naturally on the inside of the retina and where

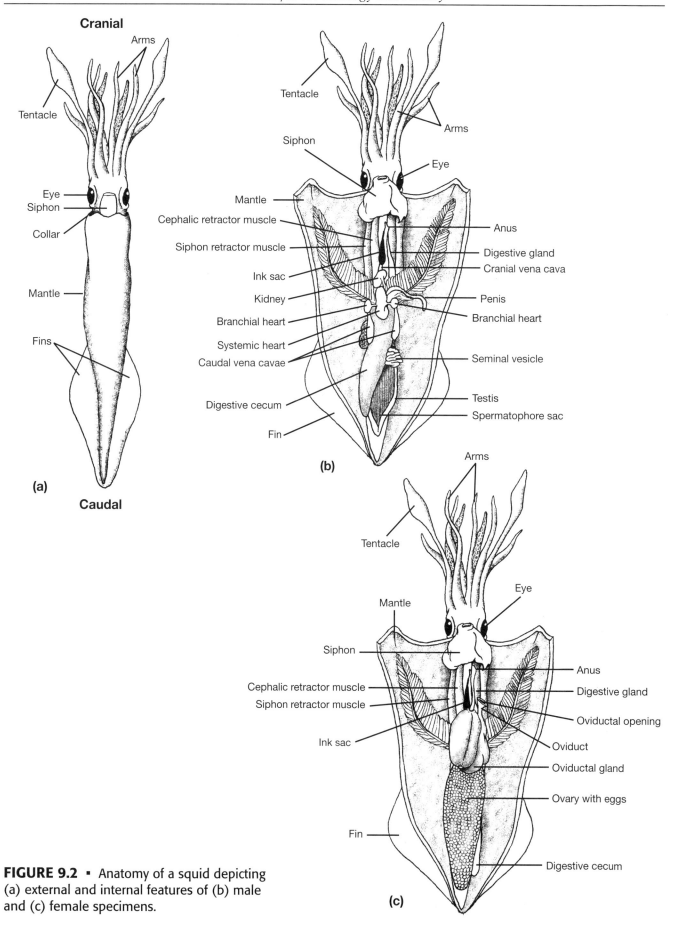

FIGURE 9.2 ▪ Anatomy of a squid depicting (a) external and internal features of (b) male and (c) female specimens.

they exit the eye as the optic nerve they come together in a large, cable-like nerve fiber and "push" the rods and cones aside to make a path through the back of the eye creating a blind spot in our visual field (Fig. 9.3a).

In contrast, the cephalopod eye has the light sensitive end of its photoreceptors oriented toward the front of the eye, so light entering the pupil passes through the lens and directly stimulates the photoreceptors. Due to the arrangement of neurons being positioned behind the retina, cephalopods have no blind spot in their visual field (Fig. 9.3b). If you think about it, this is a much more logical design for an eye than the architecture of the vertebrate eye! This is a good illustration of how different evolutionary means may be employed to reach the same end—in this case a functional, image-forming eye. Remember, natural selection can only operate on existing variation within populations, so the most logical design may not always be achievable. Without natural variability in a trait (e.g., the elasticity of the lens), that trait will forever remain unchanged and other traits in which variability within the population does exist are modified to improve survival (e.g., ciliary muscles that move the lens back and forth rather than change the shape of the lens).

TABLE 9.2 ▪ Anatomy of a Cephalopod (Squid)

Structure	Function
Collar	Fleshy border separating head-foot from visceral mass (mantle)
Eyes	Image-forming organs for detecting visual stimuli
Siphon	Hollow tube through which water is expelled from the mantle cavity at high velocity to propel the squid through the water
Mantle	Body tube encircling visceral mass forming a hollow chamber in which water is collected and used for propulsion
Arms	Shorter appendages (8) used to manipulate captured prey and act as a rudder for navigating while swimming
Tentacles	Long, extensible, prehensile appendages (2) for capturing prey
Fins	Triangular-shaped extensions of the caudal end of the body tube that are used for leisurely swimming and for maneuvering during locomotion
Gills	Feathery organs used for respiration
Esophagus	Thin tube connecting the mouth to the stomach
Stomach	Small sac located at caudal end of the body tube where food is stored and digested; digestion is entirely extracellular in cephalopods
Pancreas	Small, granular digestive gland that secretes enzymes into the stomach to assist in the breakdown of food
Liver	Large, elongated gland that releases secretions into the stomach to facilitate enzymatic digestion of food
Anus	Terminal portion of digestive tract located near siphon
Testis	Produces sperm; located in caudal end of body tube
Ink sac	Large sac that opens into the anus and secretes a dark brown or black fluid when the animal is alarmed
Siphon retractor muscles	Long muscles which control the contraction of the siphon
Kidneys	Adjacent excretory organs located between the gills
Systemic heart	Large, muscularized chamber that receives oxygenated blood from the gills and pumps it throughout the body
Branchial hearts	Smaller, muscularized chambers that receive deoxygenated blood from all parts of the body and pump blood to the gills
Caudal vena cava	Drains deoxygenated blood from the body tube and mantle back to the branchial hearts
Cranial vena cava	Drains deoxygenated blood from the head-foot back to the branchial hearts

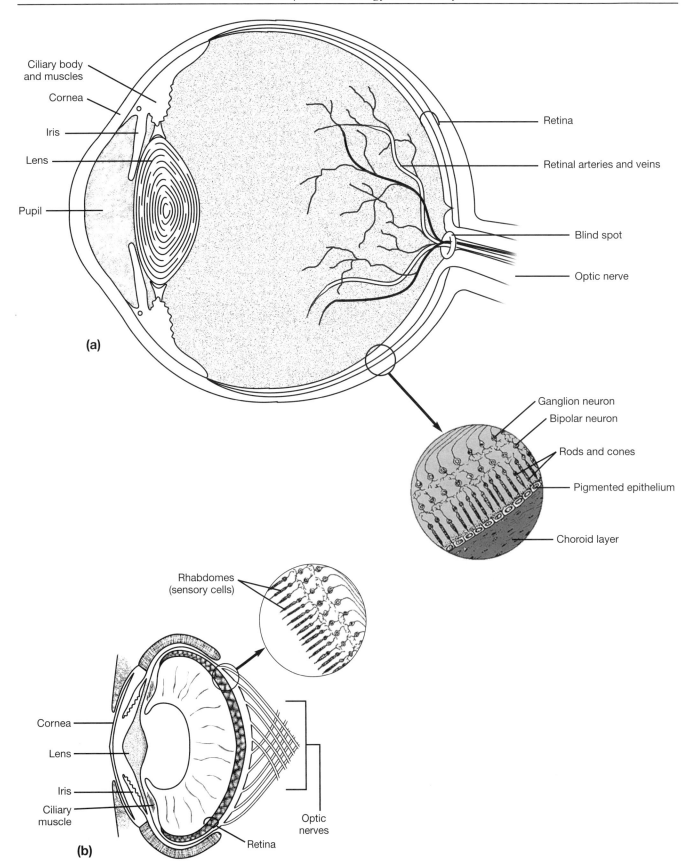

FIGURE 9.3 • Comparison of internal anatomy of (a) vertebrate eye to (b) cephalopod eye showing patterns of convergent evolution in distantly related groups.

Check Your Progress

1. Why do you think only certain tentacles possess suckers?

2. Why are sensory structures more prevalent on cephalopods than on bivalves?

Questions For Review

1. What are the four major characteristic features of molluscs?

2. How does the foot of a bivalve differ from the foot of a cephalopod?

3. In what ways do these differences reflect specific adaptations for each animal to its unique lifestyle?

4. List several features of squid which contribute to their success as predators.

5. What are some advantages of freshwater mussels producing parasitic larvae that attach to the gills of fish?

6. What is the excretory organ of bivalves?

7. Define adaptive radiation and discuss how molluscs demonstrate this phenomenon.

8. Discuss how the circulatory system of cephalopods has been adapted to suit their active, predatory lifestyle.

9. If you were an engineer hired to design an eye, would you base your model on the eye of a mammal or on that of a squid? Explain.

Annelida

After completing the exercises in this chapter, you should be able to:

1. Discuss the major characteristics of annelids.
2. Identify the major anatomical features of clamworms and earthworms.
3. Characterize patterns of locomotion of different annelids.
4. Discuss the evolutionary implications of segmentation and closed circulatory systems.
5. Define all boldface terms.

Within the phylum Annelida there are approximately 7,000 species of segmented worms divided into three major classes: Polychaeta (clamworms and sandworms), Oligochaeta (earthworms and blackworms) and Hirudinea (leeches). They occupy a wide variety of habitats ranging from marine and freshwater areas to moist terrestrial locations. Major characteristics of this phylum include a **true coelom, segmentation** (the repetition of body regions containing similar organs), a **closed circulatory system** consisting of pumping vessels (hearts), arteries and veins, a **complete digestive system** with specialized subregions, and **setae**—small, hairlike bristles used for locomotion. In fact, the degree of setal development is one distinguishing feature biologists use to divide annelids into the three separate classes. Most annelids are **hermaphroditic**,

meaning that an individual contains both male and female sex organs. They do not self-fertilize like many flatworms, however. Instead, two worms typically exchange sperm simultaneously cross-fertilizing each other's eggs.

In the following exercises you will examine several different members of this phylum commonly found in marine, freshwater and terrestrial habitats. Most of these annelids are free-living. Leeches, however, are classified as semi-parasitic since they may utilize host organisms for nutrition during part of their lives. As you examine the different members of this phylum keep in mind the many anatomical similarities that link them together and illustrate their common ancestry, yet pay close attention to the differences between each group that reflect specific adaptations to particular lifestyles.

EXERCISE 10–1	Photographic Atlas Reference Pages 53–54

Polychaete Anatomy

Materials needed:

- preserved specimen of *Nereis*
- slide of *Nereis* c.s.
- compound microscope

1. Obtain a preserved clamworm (*Nereis*) and examine its external features (Fig. 10.1). Four small **eyes** and an array of thin **tentacles** are present on the head and are part of the sensory system of *Nereis*. Most of the specialized mouthparts will be retracted within the pharyngeal region. It may be possible to pull the mouthparts out of the pharyngeal region with a small pair of forceps.

2. Examine the body of the clamworm. In polychaetes, the characteristic paired **setae** are located at the ends of feathery extensions called **parapodia.** The word polychaeta actually means "many bristles." These parapodia, while useful for locomotion, have a high degree of vascularization associated with them and are thus sites for gas exchange in this marine worm.

3. Examine a prepared slide of a cross-section through the body of *Nereis*. Locate the following organs and regions depicted in Figure 10.2: **dorsal blood vessel, intestine, coelom, parapodium, setae** (may not be visible on every slide).

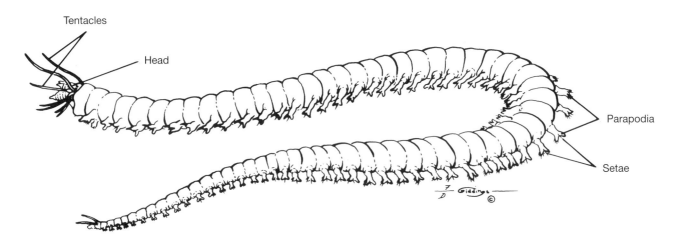

FIGURE 10.1 ▪ External anatomy of a clamworm (*Nereis*).

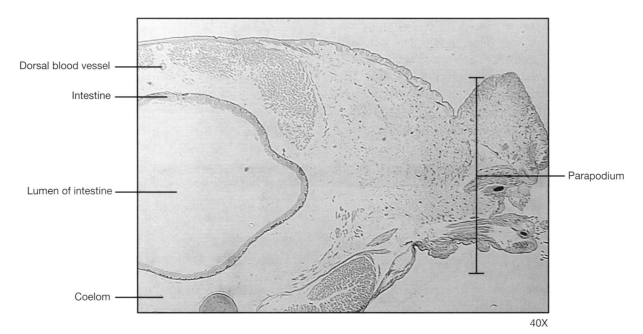

FIGURE 10.2 ▪ Cross-section through the intestinal region of a clamworm.

Check Your Progress

1. What kind of stimuli do you suppose the tentacles detect?

2. What features of polychaetes would lead you to believe that they are free-living (rather than parasitic) organisms?

EXERCISE 10–2   **Photographic Atlas Reference Pages 54–57**

Oligochaete Anatomy

Materials needed:
- preserved specimen of *Lumbricus*
- slide of *Lumbricus* c.s.
- dissection tools
- compound microscope

Members of the class Oligochaeta are free-living worms that live in freshwater habitats or damp soil. Unlike polychaetes, oligochaetes lack parapodia and rely completely on the remainder of their moist epithelial surface for gas exchange. Most possess short, bristly setae on each segment. Terrestrial members, such as the common earthworm, are well-suited for their fossorial (burrowing) lifestyle and possess many adaptive features for subterranean life.

1. Obtain a preserved specimen of *Lumbricus terrestris*, the common earthworm and examine its external anatomy. The cranial end features a small, slit-like **mouth** that is covered by the fleshy **prostomium**—an adaptation to keep dirt out of the mouth while burrowing (Fig. 10.3).

2. Run your fingers along the length of the body to feel the **setae**. These bristles help the earthworm grip the dirt and assist in locomotion.

3. Locate the **clitellum**—a large band covering several segments about one third of the way down the body from the head. The clitellum is used during reproduction for the transfer of sperm between individuals and in the secretion of a cocoon which contains the fertilized eggs.

4. At the caudal end of the body, locate the **anus**—the opening through which indigestible products are released from the digestive tract.

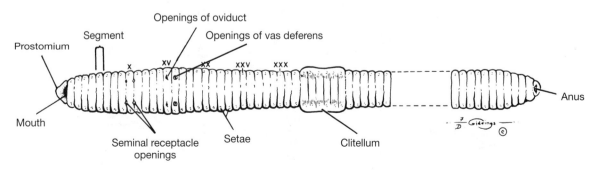

FIGURE 10.3 ▪ Ventral view of the external anatomy of the earthworm.

Check Your Progress

1. Do the setae feel different if you run your fingers in different directions along the body?

2. What does this tell you about an earthworm's ability to grip the soil?

5. Insert the tip of your dissecting scissors into the dorsal surface of the earthworm just cranial to the clitellum. Make a shallow incision progressing cranially, snipping a segment or two at a time. Remember to keep the point of your scissors pulled up against the dorsal body wall so as not to damage any internal organs.

6. Lengthen the incision all the way to the tip of the first segment (the prostomium).

7. Use pins to hold the body wall open as you identify the internal organs depicted in Figures 10.4 and 10.5. The functions of these structures are discussed in Table 10.1. *NOTE:* If you wish to use a dissecting microscope to view the internal anatomy of your earthworm, pin your specimen near one side of your dissecting pan.

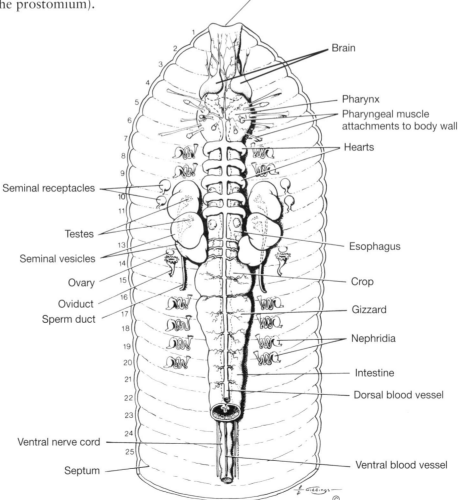

FIGURE 10.4 ▪ Dorsal dissection of the earthworm depicting internal anatomy.

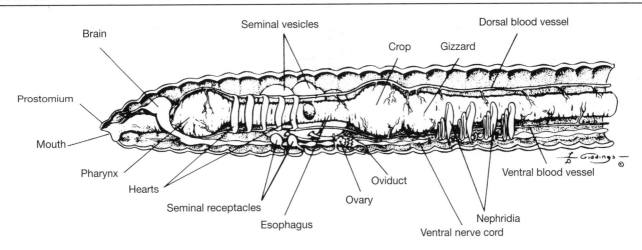

FIGURE 10.5 ▪ Lateral view of the earthworm depicting internal anatomy.

TABLE 10.1 ▪ Internal Anatomy of an Earthworm (*Lumbricus terrestris*)

Structure	Function
Mouth	Ingests soil
Pharynx	Muscularized region of digestive system specialized for pumping in soil
Esophagus	Passageway between pharynx and crop
Crop	Thin-walled chamber where food is stored
Gizzard	Thick-walled, muscularized chamber where soil is ground and usable organic materials are separated from indigestible materials
Intestine	Long tube occupying nearly two-thirds the length of the body in which nutrients are absorbed into the bloodstream
Hearts	Specialized, muscularized branches of the dorsal blood vessel that rhythmically contract to pump blood throughout the body
Dorsal blood vessel	Longitudinal blood vessel that distributes blood to the dorsal aspect of the body
Ventral blood vessel	Longitudinal blood vessel that distributes blood to the ventral aspect of the body
Seminal vesicles	Cream-colored, lobed organs fastened ventrally, but extending dorsally around each side of the esophagus that store maturing sperm
Testes (not visible)	Site of sperm production
Seminal receptacles	Ventrally located organs that receive sperm during copulation and store sperm until needed to fertilize eggs in cocoons
Ovaries	Site of egg production
Nephridia	Paired excretory organs found along the lateral margins of all but the most cranial and caudal segments; they release waste fluids out of the worm through small pores in the body wall
Septa	Thin, fleshy partitions between segments
Brain	Small, bi-lobed structure lying dorsal to the pharynx in segments 3 and 4; houses the majority of neural ganglia in the worm
Ventral nerve cord	Long, white "cord" located along the ventral surface of the body; contains large swellings of ganglia in each segment that handle the majority of coordination without intervention of the brain

Check Your Progress

1. Are both male and female reproductive structures present in your specimen?

2. List several internal organs that are exemplary of segmentation in the earthworm.

3. Put the following digestive structures in the order in which food passes through them in the earthworm: intestine, esophagus, gizzard, anus, pharynx, mouth, crop.

8. Examine a prepared slide of a cross-section through the body of an earthworm. Locate the following organs and structures depicted in Figure 10.6: **dorsal blood vessel, intestine, coelom, ventral nerve cord, epidermis, circular muscles, longitudinal muscles, nephridium, ventral blood vessel,** and **setae** (these may not be visible on every slide).

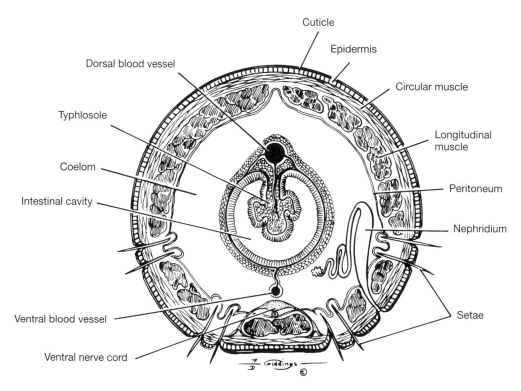

FIGURE 10.6 ▪ Cross-section through intestine of earthworm.

EXERCISE 10–3   **Photographic Atlas Reference Page 58**

Hirudinea Anatomy and Locomotion

Materials needed:
- living and preserved specimens of leeches
- prepared slide of *Hirudo* or other leech (w.m.)
- compound microscope

Leeches are members of the class Hirudinea and represent an interesting subgroup of freshwater annelids. Semi-parasitic in nature, these worms typically spend some portion of their lifecycle in close association with a host. Leeches are **ectoparasites**, meaning they attach themselves to the outside of the host organism to obtain their nourishment. This feature of leeches has been exploited by medical practitioners as far back as medieval times. Even today, doctors may attach leeches to patients to extract fluids that accumulate around injuries or to enhance the healing of incisions associated with surgeries. This may sound barbaric, but leeches are able to extract fluid much more efficiently and with less tissue damage than does hypodermic suction. These organisms can quickly consume five to ten times their original body weight in blood!

1. Obtain a prepared slide of *Hirudo* or other leech species.

2. Notice the prominent **cranial sucker** and **caudal sucker** (Figure 10.7). The **mouth** is located within the opening of the cranial sucker. The body is clearly **segmented** and you should be able to detect repeated elements within the body cavity—a defining feature of true segmentation.

3. You may be able to see outpockets of the **intestine** in each segment along with paired male and female **reproductive structures**.

4. Compare the internal anatomy of the leech with that of the clamworm and earthworm viewed earlier.

5. If living leeches are available in your laboratory, place one in a beaker of pond water and observe its movements.

6. Remove the leech from the beaker of water and place it on a wet glass plate or petri dish and observe its unique style of "creeping" when out of the water.

7. Examine the living leech with a dissecting microscope to view its suckers more closely.

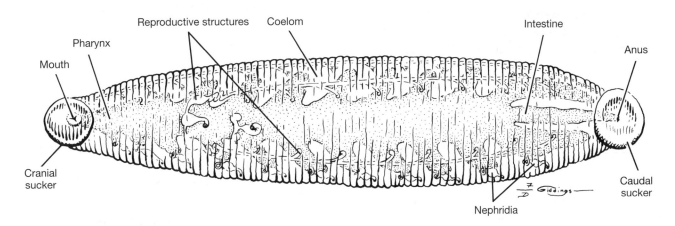

FIGURE 10.7 ▪ Anatomy of a leech.

Check Your Progress

1. What anatomical similarities do you see among leeches, earthworms and clamworms?

2. Do leeches possess setae? How do they move?

3. What is the function of the cranial sucker? the caudal sucker?

EXERCISE 10–4

Investigations with Blackworms

Materials needed:

- blackworms (*Lumbriculus variegatus*)
- spring water
- petri dishes
- filter paper
- eyedropper or disposable pipette
- stopwatch or timer
- 1" long, straight human hairs (or paint brush bristles)
- narrow rubber bands
- applicator sticks
- masking tape
- 5½" × ⅞" plastic weigh boats
- dissecting microscope

The freshwater oligochaete, *Lumbriculus variegatus*, (often known as the blackworm or mud worm) is a common occupant of the shallow margins of ponds, marshes and lakes throughout North America. Despite its wide distribution, its presence often goes unnoticed by the casual recreational user of such waters. Rarely exceeding 1–2 inches in length, this small, free-living annelid has proven to be easily obtained and cultured in the laboratory and can be used to illustrate a host of biological principles including: patterned regeneration of lost body parts, blood vessel pulsations, swimming and other locomotor reflexes, and giant nerve fiber action potentials.

OBSERVATION OF CIRCULATION

Annelids possess a **closed circulatory system** consisting of muscularized pumping vessels that channel blood through a contained series of arteries and veins. In this exercise you will see the closed circulatory system of the blackworm in action. In *L. variegatus* the blood is pumped by rhythmic contractions of the dorsal blood vessel, moving the blood from the caudal end (where gas exchange occurs) toward the head. The blood is red and thus easy to distinguish from other body fluids due to the respiratory pigment **erythrocruorin**, a hemoglobin-like pigment that reversibly binds oxygen and carbon dioxide in much the same way that hemoglobin works in humans.

1. Use an eyedropper or disposable pipette to place a worm on moist filter paper.

NOTE: Exposure to chlorinated water or soap residues will quickly kill these worms. Be sure you are using de-chlorinated water or other invertebrate-safe culture and clean, chemical-free glassware and pipettes.

2. Head and tail regions are easy to distinguish because the head segments are blunter, wider, more maneuverable and more darkly pigmented than tail segments. In addition, the prostomium, muscular pharynx and reproductive organs are contained within the first 8–10 cranial segments.

3. Position your specimen under the dissecting microscope and be sure you can differentiate between the head and tail.

4. You will notice two large blood vessels running longitudinally along the body. These are the dorsal and ventral blood vessels. The **dorsal blood vessel** is contractile and is responsible for the distribution of blood through the closed circulatory system. The **ventral blood vessel**, although large and easily visible, is not contractile.

5. Count the number of pulsations of the dorsal blood vessel for 1 minute and record your finding below.

 # of pulsations/minute _____

6. Closer inspection of the circulatory system should reveal numerous lateral branches of the dorsal blood vessel that transport blood to the organs and tissues within each segment.

7. In nature, the blackworm uses its head to forage for vegetation and microorganisms while its tail end projects upwards and bends at a right angle, just breaking the surface tension of the water. This posture allows gases to be exchanged between the air and the pulsating dorsal blood vessel lying just below the body surface. Examine the dorsal blood vessel in the tail region of your specimen. Do you notice any anatomical or physiological adaptations in this region that would enhance gas exchange?

Check Your Progress

1. How did the pulsation rate of your specimen compare to others in your laboratory?

2. Collect pulsation rates from at least 5 other groups and calculate a mean pulsation rate for blackworms in your laboratory.

3. What factors do you suspect could affect the pulsation rate of the dorsal blood vessel in blackworms?

OBSERVATION OF LOCOMOTION AND REFLEXES

No doubt you have all seen an earthworm use peristaltic crawling to move across a sidewalk after a heavy rain or quickly recoil into its burrow when disturbed. Earlier in this chapter you had the opportunity to observe the unique locomotor patterns of another annelid, the leech. In this exercise you will see how blackworms utilize these same circular and longitudinal muscle layers to accomplish a surprisingly diverse array of acrobatic locomotor movements.

1. First assemble a tactile probe. Cut a narrow rubber band into a 1" long, straight section and tape it to the end of a wooden applicator stick

so that approximately 1/2" of the rubber band protrudes past the end of the applicator stick.

2. Fashion another tactile probe out of a 1" long, coarse, straight human hair or paint brush bristle using the same process as described in Step #1.

3. Place a blackworm on moist filter paper.

4. Using the applicator stick with the rubber band section, lightly stroke the worm's tail.

5. Record your observations in Table 10.2.

6. Next lightly stroke the worm's head using the same tactile probe and record your observations.

7. Determine whether a weak stimulus (the tactile probe with the hair) is as effective as the strong stimulus (the tactile probe with the rubber band) in evoking crawling behavior.

8. In the space provided in Table 10.2 make a sketch of the worm and label the head and tail ends. On the sketch, depict all locations along the body where touching elicits *forward* crawling. Indicate where it fails to do so. Try touching at least 6–8 different locations along the body, moving progressively forward from the tail to the head. The result is a map of the touch sensory field for forward crawling.

9. Repeat this process to generate a touch sensory field map for rearward crawling by starting at the head and working backward.

10. Construct two more tactile probes using 2" rubber band and hair sections this time. Instead of leaving one end free, as before, form a loop with the rubber band and hair sections and tape the two free ends to the wooden applicator sticks. In each case, this will leave a loop approximately 1/2" long.

11. Transfer a fresh, unused worm to a plastic weigh boat containing 1–2 cm of spring water.

NOTE: For added strength, you may want to stack several weigh boats together. Clear petri dishes with white paper underneath may be used in place of weigh boats.

12. Coax the worm to the center of the container using the tactile probe with the rubber loop.

13. After the worm has settled there, hold the tactile probe (with the rubber loop) at a 45° angle and quickly press the loop down on the tail for a brief instant. If done properly, this should not injure or fragment the worm but, instead, induce swimming.

14. Record your observations in Table 10.3.

15. Repeat Step #11, but this time press gently on the worm's head. Record your observations. In this case, the stimulus should not elicit swimming, but rather a reversal response in which the worm quickly curls and then uncurls its body, causing a nearly 180° repositioning of its head and tail. Since blackworms cannot swim backwards, reversal enables rapid repositioning of the head, presumably moving it away from the stimulus or threat and making subsequent forward swimming possible.

16. As before, determine whether a weak stimulus (the probe with the hair loop) is as effective in evoking swimming or reversal behavior from both ends.

TABLE 10.2 ▪ Locomotor Properties of Blackworms in Moist, Terrestrial Environments

Treatment	Stimulus	Movement evoked?	Direction of movement	Description of movement
Tail stroked	Strong			
Head stroked	Strong			
Tail stroked	Weak			
Head stroked	Weak			
Touch sensory field map for forward crawling				
Touch sensory field map for rearward crawling				

17. In the space provided in Table 10.3 make a sketch of the worm and label the head and tail ends. On the sketch, depict all locations along the body where touching elicits *forward swimming*. Indicate where it fails to do so. As before, try touching at least 6–8 different locations along the body, moving progressively forward from the tail to the head to construct a map of the touch sensory field for forward swimming.

18. Repeat this process to generate a touch sensory field map for reversal behavior by starting at the head and working backward.

TABLE 10.3 ▪ Locomotor Properties of Blackworms in Underwater Environments

Treatment	Stimulus	Movement evoked?	Direction of movement	Description of movement
Tail prodded	Strong			
Head prodded	Strong			
Tail prodded	Weak			
Head prodded	Weak			
Touch sensory field map for forward swimming				
Touch sensory field map for reversal behavior				

Check Your Progress

1. In both environments, were strong stimuli more effective at evoking locomotor behaviors than weak stimuli?

2. Was peristaltic movement employed by blackworms for crawling? Is this more similar to locomotion in earthworms or leeches?

3. Was peristaltic movement employed by blackworms for swimming?

4. How would you describe the movements employed by blackworms for swimming?

Questions For Review

1. List 3 major distinguishing features of annelids.

2. What is a closed circulatory system?

3. How do the nephridia of annelids compare to the flame cells of flatworms?

4. Why must terrestrial annelids such as earthworms maintain a moist skin surface?

5. Describe the advancements in the specialization of the digestive tract in annelids in comparison to the digestive tract of *Ascaris* (nematode).

6. Discuss the difference in the role of the cuticle in the nematode (*Ascaris*) and the earthworm (*Lumbricus*).

7. List two major features that nematodes and earthworms share in common **and** two that are found in earthworms but not in nematodes.

   Nematodes and Earthworms: **Earthworms ONLY:**

   1.   1.

   2.   2.

8. Discuss the differences between (1) fertilization in dioecious animals such as mammals, (2) self-fertilization in monoecious animals such as tapeworms and (3) cross-fertilization in hermaphroditic (monoecious) animals such as earthworms.

Arthropoda

After completing the exercises in this chapter, you should be able to:

1. Identify the internal and external anatomy of a crayfish and discuss the functions of each structure.

2. Identify the internal and external anatomy of a grasshopper and discuss the functions of each structure.

3. Categorize orientation behaviors in animals and discuss the evolutionary significance and possible origins of such behaviors.

4. Define all boldface terms.

The phylum Arthropoda is by far the largest in the animal kingdom, containing an estimated 10 million species! New species of arthropods are literally being discovered every day, adding to the nearly one million which have already been described. As their numbers suggest, they are perhaps the most successful group of animals ever to occupy the planet. They predate dinosaurs (by several hundred million years!) and most surely will be feeding upon the last vertebrate corpse as it slowly decays. Part of their unsurpassed success is due to the fact that they were the first animals to inhabit land. Between 440 and 410 million years ago, arthropods gradually moved into a previously unexploited habitat that simultaneously was being populated by vascular plants. They were also the first animal group to evolve the ability to fly and therefore could make use of the 3-dimensional landscape that was devoid of any other competitors. This resulted in an adaptive radiation of arthropods throughout the landscape. These facts, coupled with their small size, made them ideal vectors of pollen and explain the close associations we see today between many flowering plants and insects. Throughout their long evolutionary history, arthropods have spread through every terrestrial, aquatic and aerial habitat imaginable and have had a profound impact on the evolution of numerous other species.

Despite their diversity, all arthropods share many characteristics in common. Perhaps the most universal traits of arthropods are the presence of a **segmented body** and **jointed appendages**, the latter trait from which the phylum name is derived. Arthropods also possess a hard, chitinous **exoskeleton** that is secreted by the epidermis. As the body grows inside, the old exoskeleton is periodically shed through a process called **molting** in which the soft, new exoskeleton must be secreted and fixed in place before the old shell can be shed. This causes severe logistic problems (as you may imagine) that arthropods circumvent by folding the soft, new exoskeleton upon itself as it is being produced. After the old shell is shed and the body is free of its former constraints, the new, larger shell expands to its final size and hardens in place. In an animal encased in such a rigid suit of armor, the coelom can play no major role in locomotion and is thus a greatly **reduced coelom**. The main body cavity is instead a hemocoel, comprising part of the **open circulatory system** characteristic of this group.

Today, three classes of arthropods (Arachnida, Insecta and Malacostraca) include well in excess of 95% of all arthropod species. Providing a complete survey of the arthropod phylum would be an impossible task, however most adult arthropods show only minor deviations from the "standard" body plan. Thus the crayfish and grasshopper are presented as representative body styles of the typical aquatic and terrestrial arthropod.

Crayfish Anatomy

Materials needed:
- preserved crayfish (*Cambarus*)
- dissecting tools
- dissecting pan
- dissecting microscope

EXTERNAL ANATOMY

The crayfish is a member of the Class Malacostraca (formerly Crustacea) which includes lobsters, shrimps and crabs. The crayfish is one of the few freshwater members of this group and serves as an excellent model for studying the magnificent adaptations that aquatic arthropods have developed. Two defining features of malacostracans are their **biramous** (Y-shaped) appendages and their two sets of **antennae**.

1. Obtain a preserved specimen of *Cambarus*, the freshwater crayfish.

2. The body is divided into two main regions: the **cephalothorax** and the **abdomen**. Notice the segmented nature of these regions and the many modified appendages present in each of these areas (Fig. 11.1).

3. Examine the external anatomy of the crayfish and identify the structures listed in Table 11.1.

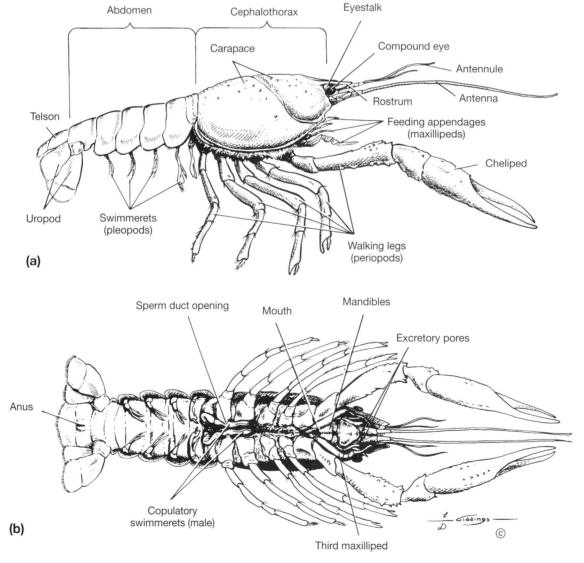

FIGURE 11.1 · External anatomy of the crayfish (*Cambarus*): (a) lateral view and (b) ventral view.

TABLE 11.1 ▪ External Anatomy of the Crayfish

Structure	Function
Chelipeds	Large pinchers used for grasping food and for defense
Walking legs (periopods)	Locomotion (walking on land and crawling across stream bottom)
Swimmerets (pleopods)	Modified caudal appendages for swimming
Copulatory swimmerets (male)	Larger, club-shaped swimmerets used by male to stimulate female during copulation and fertilization
Uropod and telson (tail)	Broad, fan-shaped region of body used for rapid movement and for directional control during leisurely locomotion
Antennae	Longer, paired appendages on head modified for chemosensory and tactile reception
Antennules	Shorter, paired appendages on head modified for chemosensory and tactile reception
Compound eyes	Small, dark sense organs for detecting light and forming visual images
Rostrum	Pointed region between eyes demarcating cranial end of body
Maxillipeds	Three sets of paired appendages located on ventral surface near mouth used to manipulate food
Mandibles	Hard, chitinous mouthparts used to manipulate food into mouth
Mouth	Opening to digestive tract located on ventral surface of body
Anus	Terminal point of digestive system located on ventral surface of uropod

Check Your Progress

1. What advantage might mouthparts of various shapes provide a crayfish?

2. Are the antennae biramous? Are the antennules biramous?

3. What two "typical" arthropod body regions are fused into the cephalothorax?

4. Is your crayfish male or female? How do you know?

INTERNAL ANATOMY

1. Using the pointed end of your dissecting scissors, make two incisions along the dorso-lateral margins of the crayfish from the caudal end of the cephalothorax to the rostrum, angling the incisions medially so they meet at the tip of the rostrum.

2. Gently peel away the dorsal portion of the carapace, being careful not to pull away any of the internal organs that are attached to the underside of the shell.

3. Next, remove one of the remaining lateral sides of the carapace to reveal the feathery, branched **gills**.

4. Notice that the gills are actually external. They reside between two pieces of exoskeleton—the outer lateral side of the carapace and a thinner, inner chitinous membrane.

5. Notice also that the gills are attached to the walking legs at their ventral juncture.

6. Continue the dorso-lateral incisions made earlier, this time directing them caudally through the abdomen toward the telson.

7. Carefully remove the portion of the exoskeleton covering the dorsal surface of the abdomen to expose the musculature of the tail and the blood vessels and digestive organs located within this region.

8. Work from the dorsal surface downward (ventrally), identifying the organs and structures described in Table 11.2 and depicted in Figure 11.2 as you go.

9. After differentiating between the **cardiac chamber** and the **pyloric chamber** of the stomach, remove them from the crayfish and open the cardiac chamber to reveal the **gastric mill**. The crayfish uses these chitinous teeth to mechanically grind its food into smaller pieces for digestive enzymes secreted by the **digestive glands** to act upon as the food moves into the pyloric chamber of the stomach.

10. Use a dissecting microscope to scan the dorsal aspect of the digestive glands for the small reproductive structures. **Ovaries** will be easier to spot than the extremely small **testes**, but careful

TABLE 11.2 ▪ Internal Anatomy of the Crayfish

Structure	Function
Gills	Respiration
Esophagus	Passageway between mouth and cardiac portion of the stomach
Cardiac chamber of stomach	Thick-walled, cranial portion of the stomach containing gastric mill—chitinous teeth which grind food into smaller pieces
Pyloric chamber of stomach	Thin-walled chamber where chemical digestion of food occurs
Digestive glands	Accessory digestive organs that secrete enzymes into the pyloric stomach to facilitate chemical breakdown of the food
Intestine	Long tube passing through the "tail" region in which nutrients are absorbed into the bloodstream for delivery to the body tissues
Heart	Specialized, muscularized chamber containing ostia (holes) to allow passive uptake of blood which is delivered to the body tissues through arteries, but veins do not exist in this open system
Dorsal artery	Longitudinal blood vessel that distributes blood to the dorsal aspect of the body
Ventral artery	Longitudinal blood vessel that distributes blood to the ventral aspect of the body
Testes (male)	Site of sperm production (may be difficult to locate on specimen)
Ovaries (female)	Site of egg production
Green glands	Paired excretory organs found along the ventral margin of the head region; they release waste out of the crayfish through small pores in the ventral body wall
Brain	Small, radiate structure lying dorsal to the green glands; houses the majority of neural ganglia in the crayfish
Circumesophageal connection (of ventral nerve cord)	Branches of the ventral nerve cord that bifurcate at the base of the brain and wrap around the esophagus before merging along the ventral surface of the crayfish just caudal to the esophagus
Ventral nerve cord	Long, white "cord" located along the ventral surface of the body; contains large swellings of ganglia that handle the majority of coordination without intervention by the brain

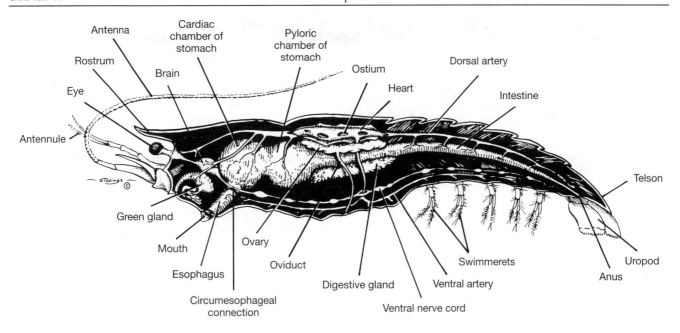

FIGURE 11.2 ▪ Lateral view of the internal anatomy of a female crayfish.

examination of this region along the median plane of the crayfish should reveal the reproductive organs. The size of the testes and ovaries and the presence of eggs within the ovaries depends upon the time of year that the specimens were collected, since reproduction in this group is seasonal.

Check Your Progress

1. What structure in the cardiac chamber of the stomach allows crayfish to mechanically digest their food?

2. What organ secretes enzymes into the pyloric chamber of the stomach to assist in the chemical digestion of food?

3. How do the relative positions of the brain and nerve cord in crayfish compare to the positions of the brain and nerve cord in the earthworm? Is this evidence of common ancestry?

4. Does the crayfish heart possess holes (openings)? What type of circulatory system does this represent?

5. Where on the body surface of the crayfish are the external openings for the excretory organs (green glands) located?

Grasshopper Anatomy

Materials needed:
- preserved grasshopper (*Romalea*)
- dissecting tools
- dissecting pan
- dissecting microscope

The grasshopper serves as a representative example of the class Insecta, a subgroup of arthropods with six, **uniramous** (unbranched) walking appendages, a single set of **antennae** and three distinct body regions: **head, thorax** and **abdomen.** During the evolution of arthropods, departure from the ancestral aquatic lifestyle favored the development of characteristics that permitted successful adaptations to the many ecological hurdles associated with terrestrial living. Among them, (1) stronger, more efficient support systems and walking appendages, (2) waxy **cuticles** built to withstand the osmotic stresses of "dry" air, yet permeable enough for aerial gas exchange, (3) the ability to fertilize eggs internally to prevent desiccation of "naked" gametes, (4) specialized excretory and digestive structures designed to conserve water and (5) appendages modified into **wings** to take advantage of the previously unexploited aerial habitat, were hallmarks in the evolution of insects.

EXTERNAL ANATOMY

1. Obtain a preserved specimen of *Romalea*, the common grasshopper.

2. The body is divided into three main regions: the **head,** the **thorax** and the **abdomen.** Notice the segmented nature of these regions and the many modified appendages present in each of these areas (Figure 11.3).

3. Determining the sex of your grasshopper can be accomplished by examining the caudal portion of the abdomen. In addition to an anus, females possess an **ovipositor** ventral to the anus. This opening is bordered by two pairs of chitinous "teeth" that thrust into the soil and flex outward, creating a chamber in which the female deposits her eggs. Males lack an ovipositor.

4. Examine the external anatomy of the grasshopper and identify the structures listed in Table 11.3.

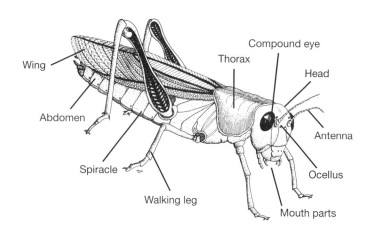

FIGURE 11.3 ▪ External anatomy of the grasshopper (*Romalea*).

TABLE 11.3 ▪ External Anatomy of the Grasshopper

Structure	Function
Spiracles	External openings in abdomen that allow air flow into and out of tracheae
Compound eyes	Paired, complex, image-forming photoreceptors composed of numerous ommatidia which create a fairly coarse-grained picture of their visual field
Ocellus (pl. = ocelli)	Simple photoreceptor consisting of a small cup backed by light-absorbing pigments; not capable of image formation
Antennae	Thin, paired appendages on head modified for chemosensory and tactile reception
Mouthparts	Multiple sets of hard, chitinous appendages used to chew food and manipulate food into mouth
Wings	Two sets of paired appendages modified for flight

Check Your Progress

1. Are the grasshopper's antennae biramous?

2. How many major body regions do insects have?

3. How many pairs of walking appendages do insects possess?

4. What other appendages are present for locomotion?

5. Is your grasshopper male or female? How do you know?

6. Why do the spiracles of the respiratory tract open to the outside of the body?

INTERNAL ANATOMY

1. Using the pointed end of your dissecting scissors, make an incision along the dorsal margin of the grasshopper from the caudal end of the abdomen to the head.

2. Gently peel apart the lateral portions of the thorax and abdomen, being careful not to pull away any of the internal organs that are attached to the underside of the exoskeleton.

3. Notice the thin, filamentous tubules stretching between the internal organs and the **spiracles** which open on the outside of the abdomen. These are **tracheae**—respiratory tubules of the grasshopper that conduct air flow from the outside environment directly to the tissues within the body.

4. In grasshoppers, **ovaries** will be easier to spot than the small **testes**, but careful examination of the dorsal region along the median plane of the specimen should reveal the reproductive organs. The size of the testes and ovaries and the presence of eggs within the ovaries depends upon the time of year that the specimens were collected, since reproduction in grasshoppers is seasonal.

5. Work from the dorsal surface downward (ventrally), identifying the organs and structures described in Table 11.4 and depicted in Figure 11.4 as you proceed.

TABLE 11.4 ▪ Internal Anatomy of the Grasshopper

Structure	Function
Esophagus	Passageway between mouth and crop
Crop	Highly extensible, cranial portion of the digestive tract that serves as a storage compartment for food
Stomach	Chamber which receives semi-digested food from the crop and secretes enzymes for chemical digestion
Gastric ceca	Lobed, accessory digestive areas located at the juncture between the crop and stomach which facilitate chemical digestion
Intestine	Shortened tube passing through the caudal portion of the abdomen in which nutrients are absorbed into the bloodstream
Rectum	Specialized swelling of the caudal portion of the digestive tract for efficient water reabsorption and conservation
Anus	Terminal portion of the digestive tract which regulates egestion of undigested food from the body
Ovipositor (female)	External opening bordered by pointy, chitinous teeth that penetrate the soil and create burrows for egg deposition
Heart	Specialized, muscular swellings of the dorsal artery containing ostia (holes) to allow passive uptake of blood which is delivered to the body tissues through arteries, but veins do not exist in this open system
Testes (male)	Site of sperm production (may not be visible on specimen)
Ovaries (female)	Site of egg production and maturation; fertilization is internal and fertilized eggs are deposited in the soil
Malpighian tubules	Stringy, fibrous excretory organs scattered along the margins of the stomach and intestine which release metabolic waste into the digestive tract so that excess water may be reabsorbed by the intestine and rectum and conserved in the body
Brain	Enlarged conglomeration of nervous tissue located in the head that processes sensory information and coordinates major body functions
Ventral nerve cord	Long, whitish "cord" located along the ventral surface of the body; contains large swellings of ganglia that handle some nervous coordination without intervention by the brain

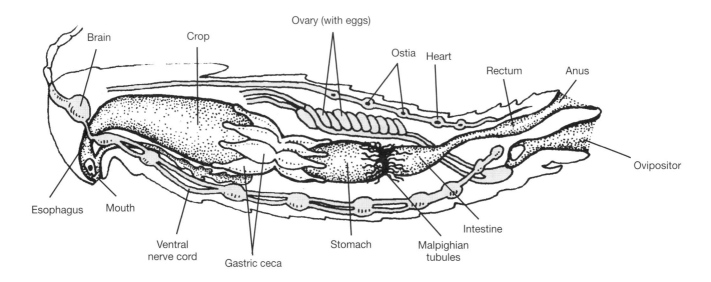

FIGURE 11.4 ▪ Lateral view of the internal anatomy of a female grasshopper.

Check Your Progress

1. What digestive organ for food storage do grasshoppers possess that bears the same name in earthworms?

2. What specialized area of the digestive tract reflects an adaptation to living in dry, terrestrial habitats?

3. What organ in grasshoppers performs the same function as the green glands in crayfish?

4. How does the position of the nerve cord in the body of grasshoppers compare to its position in crayfish?

5. What benefit do the gastric ceca provide for grasshoppers?

EXERCISE 11–3

Orientation Behaviors

Behavior can be defined generally as any overt response by an animal to a stimulus; thus we can say that **orientation behaviors** refer to specific responses that involve positioning or movements by an animal.

A natural starting point in a behavioral study is to attempt to elucidate the nature of the stimulus that evokes a particular behavior. In other words, we typically observe the behavior first and then ask the question, "What causes that behavior?" Again we return to our earlier discussion of proximate versus ultimate behavioral causes from Chapter 6. If we are interested in the stimulus that elicited the behavior then we are looking for the proximate cause of the behavior. (You may also want to know *why* this behavior evolved in the first place and persisted over time in this species—the ultimate cause.) For most behaviors, light, sound, gravity, temperature, moisture or chemicals are likely candidates for potential proximate cues. Since many animals orient to stimuli for which humans have no capacity to sense (i.e., ultraviolet light or low frequency vibrations), it is often not the stimulus most obvious to us that evokes the behavior we observe.

To effectively tease apart the stimuli causing a behavior, we must first be able to precisely describe the behavior we are observing. Several different categories have been defined to include the numerous behaviors we commonly observe in animals. With respect to orientation behaviors, two main categories exist. **Kinesis** is a general term for a random behavior that does not necessarily orient the animal toward a stimulus, but rather affects its general rate of movement (orthokinesis) or degree of turning (klinokinesis). For example, terrestrial isopods respond to increases in humidity by decreasing their overall movements. This behavioral response tends to keep them in damper areas. As the humidity in their environment decreases, their rate of movement increases, presumably to carry them to a more favorable location. In contrast to kinesis, a **taxis** is a directed behavior in which an animal orients or moves toward or away from a stimulus. Both kineses and taxes may be **positive** (an increase in movement or a movement toward the stimulus) or **negative** (a decrease in movement or movement away from the stimulus). Similarly a prefix may be attached to the taxis or kinesis to further describe the stimulus, as mentioned earlier with klinokinesis and orthokinesis. Other examples of descriptive prefixes are: photo- (light), chemo- (chemical), geo- (gravity), thigmo- (touch) and thermo- (heat).

RESPONSES TO HUMIDITY AND LIGHT IN ISOPODS

Materials needed:
- terrestrial isopods (any species)
- petri dish "choice chambers"
- black cloth or construction paper
- plastic screen mesh
- paper towels
- Dririte® desiccant
- gooseneck lamp with 60- and 150-watt bulbs

Isopods are the terrestrial crustaceans often referred to as sowbugs, pillbugs or rolie-polies. Common to wooded and urban settings alike, they are generally found in leaf litter or under other debris where they feed on decaying organic matter. Isopods possess two pairs of antennae, one pair of compound eyes and seven pairs of walking legs. In the following experiment, modified petri dishes will be used to give these animals a choice between different environments, which you will manipulate. In the first part of the experiment, you will examine the preferences of isopods for humid and arid environments. After that you will examine their preferences for light and dark environments. After you have gained some knowledge about their responses to moisture and light as separate stimuli, you will examine the interplay between these two stimuli to determine which is the more powerful stimulus.

NOTE: Keep the room lights low during these laboratory exercises. Stray light may cause unexpected results in these experiments.

Experimental Protocol

1. Working in groups of 3–4 students, obtain an isopod choice chamber that has been constructed by taping two petri dish bottoms together leaving a cut-out passageway between them.

2. Place a folded, moistened paper towel in the bottom of one petri dish and sprinkle Dririte® (a commercial desiccant) in the bottom of the other half of the chamber.

3. Position the plastic screen mesh over the bottom of the two dishes and place the lids securely on the chambers. The isopods will be crawling on the top surface of the screen mesh, so be sure there are no gaps present at the edges of the chamber for escape. Rubber bands may be used to tightly secure the lids.

4. Allow the dishes to sit undisturbed for about 5 minutes to allow the humidity levels in the two chambers to adjust.

5. After 5 minutes, introduce 10 isopods in each dish through the central stoppered holes.

6. Leave the chamber undisturbed for at least 10 minutes in a setting where it is exposed to low, uniform light, preferably away from the immediate area in which you are working.

7. While the animals are "selecting" their preferred location within the dishes, answer the following questions.

Using the Scientific Method

1. What is the general question being addressed by this experiment?

2. Record the hypothesis you wish to test.

3. Identify the independent variable in this experiment.

4. Identify the dependent variable in this experiment.

5. What are the levels of treatment (i.e., levels of the independent variable)?

6. Did you set up a control treatment? If not, describe an appropriate control treatment for this experiment.

7. Identify several controlled variables.

8. After 10 minutes, return to your choice chamber and record the number of isopods in each dish. Record your findings in Table 11.5.

9. Remove the individuals from the chamber (your instructor may have a designated container for "used" isopods) and discard the Dririte® from the one dish.

10. Set up the choice chamber with moist, folded paper towels in each side of the dish, replace the lids and rubber stoppers and again let the dishes sit for 5 minutes.

TABLE 11.5 ▪ Responses to Humidity and Light in Terrestrial Isopods

Treatment	Number Found in Location
Dry	
Humid	
Light	
Dark	
Light, humid	
Dark, dry	

11. Place 10 individuals in each dish, as before, then replace the stoppers and cover the top and sides of *one of the dishes* with black cloth or construction paper. Let the choice chamber sit undisturbed for at least 10 minutes.

Using the Scientific Method

1. What is the general question being addressed by this experiment?

2. Record your hypothesis.

3. Identify the independent variable in this experiment.

4. Identify the dependent variable in this experiment.

5. What are the levels of treatment (i.e., levels of the independent variable)?

12. After 10 minutes, return to your choice chamber and record the number of isopods in each dish. Record your findings in Table 11.5.

13. Finally, design and conduct an experiment to examine the interplay between these two stimuli. Determine the preference of isopods between a brightly illuminated, humid environment and a dark, dry environment. Record your findings in Table 11.5.

Using the Scientific Method

1. Record your hypothesis.

2. What are your two independent variables in this experiment?

Check Your Progress

1. Could the isopods differentiate between the moist and dry microhabitats?

2. Based on these findings, would you retain or reject your hypothesis?

3. Summarize your findings about the preference of isopods for humid versus arid microhabitats.

4. Explain how this behavior would be adaptive for these organisms.

5. Could the isopods differentiate between the light and dark regions?

6. Summarize your findings about the preference of isopods for light versus dark environments.

7. Explain how this behavior would be adaptive for these organisms.

8. Which stimulus, light intensity or relative humidity, is more important in affecting isopod habitat selection?

9. Propose additional studies that might be performed on isopods to delve further into these or other orientation behaviors.

PHOTOTAXIS IN WATER FLEAS

Water fleas are aquatic crustaceans commonly found in freshwater ponds and lakes. Individuals possess branched antennae, a single, median compound eye and flattened legs that are their primary respiratory organs. Many zooplankton are known to make daily vertical migrations within lakes in response to changing light intensity levels, moving up within the water column after sundown and returning to greater depths as day approaches. Presumably, these vertical migrations place the zooplankton within the upper, more productive parts of the aquatic ecosystem at a time when predation pressures and damaging illumination levels are lowest. In addition to light intensity, the proportion of different wavelengths within the light may also serve as an orientation stimulus for *Daphnia*. Since water absorbs more long than short wavelengths of light, at greater depths the proportion of blue light increases while the proportion of red light decreases. Perhaps *Daphnia* use the relative difference in the frequencies of red and blue light to determine their preferred depth. Thirdly, phytoplankton contain chlorophyll (a green photosynthetic pigment) that reflects green light, while absorbing red and blue wavelengths. Perhaps *Daphnia* use this information to determine where phytoplankton populations are located in the water column. In the following exercise, you will determine the phototactic responses of *Daphnia magna* to changes in light source position, light intensity and wavelength.

Materials needed:
- live *Daphnia magna*
- 100 mL graduated cylinders
- spring water
- fine mesh net or pipettes for dispensing *Daphnia*
- ring stands w/clamps to hold cylinders and beaker plates to hold heat shields
- medium-sized petri dishes (heat shields)
- black construction paper sleeves
- colored filters: red, blue and green
- gooseneck lamps with 60-watt and 150-watt bulbs
- fluorescent lamps with 25-watt bulbs

Experimental Protocol

1. Working in small groups of 3–4 students, obtain a 100 mL graduated cylinder containing a small population of *Daphnia*.

2. Suspend the cylinder by a clamp from a ring stand and use a gooseneck lamp with a 60-watt bulb to present light first from above for several minutes, noting *Daphnia's* response. Use the ten 10-mL divisions on the cylinder to characterize the vertical distribution of the population.

3. To protect the organisms from excessive heat, use a petri dish filled with water as a heat shield. Suspend it between the cylinder and the lamp using a beaker holder.

4. Next, position the light below the cylinder and look for changes in the population's distribution. In the space provided below, describe the response of *Daphnia* to low-intensity light. Would you characterize their behavior as positively phototactic?

With these original observations in mind, you will now perform a series of experiments in which different intensities and wavelengths of light will be used. In the first experiment, you will compare the vertical migrations of *Daphnia* under controlled conditions in five treatments: (1) no light, (2) low-intensity light from above, (3) low-intensity light from below, (4) high-intensity light from above and (5) high-intensity light from below.

Using the Scientific Method

1. What is the general question being addressed by this experiment?

2. Record the hypothesis you wish to test.

3. Identify the independent variable in this experiment.

4. Identify the dependent variable in this experiment.

5. What are the levels of treatment (i.e., levels of the independent variable)?

6. Explain how your original observations constitute a control treatment.

7. Identify several controlled variables in this study.

NOTE: To expedite data collection for this experiment, set up all five treatments simultaneously but in staggered fashion, a minute or so apart. Also, each group member should be responsible for counting *Daphnia* in a specific volume of the water column, since data collection will need to be completed quickly and accurately.

5. Set up five 100 mL graduated cylinders with populations of *Daphnia* as described earlier in Steps #1 and #2.

6. To eliminate side light in this study, place sleeves of black paper around the graduated cylinders. To create the "no light" treatment, additional pieces of black paper can be applied to the top and bottom of the designated cylinder.

7. Use gooseneck lamps with 60-watt bulbs for the low-intensity light treatments and lamps with 150-watt bulbs for the high-intensity treatments. Be certain that you use heat shields for these four treatments and that you position all four lights the same distance from their respective cylinders.

8. Allow the organisms 5 minutes to adjust their positions within the water column in each treatment.

9. After 5 minutes, remove the paper sleeve from each cylinder and quickly count the number of *Daphnia* in each 10 mL volume within the cylinder. Record your data in Table 11.6.

TABLE 11.6 · Number of *Daphnia* in Each Volume of the Water Column After 5 Minutes of Exposure to Different Light Intensities

10 mL Volumes in Graduated Cylinder	No Light	Low-intensity Light		High-intensity Light	
		Light From Above	Light From Below	Light From Above	Light From Below
90–100					
80–90					
70–80					
60–70					
50–60					
40–50					
30–40					
20–30					
10–20					
0–10					

Check Your Progress

1. How does the positioning of the light from above and below allow you to tease apart the response of *Daphnia* to light from their response to gravity?

2. Did their response differ between the no-light, high-intensity light and low-intensity light treatments? How so?

3. Based on these findings, would you retain or reject your hypothesis?

Next, you will examine the affect of variations in wavelength on vertical migration patterns in *Daphnia*. One concern in using colored filters is that they do not all transmit the same light intensity. Light intensity must be the same under all conditions or the study would have two confounding treatments. Another concern is that incandescent bulbs do not contain the full spectrum of visible light wavelengths. Therefore you must use a combination of incandescent and fluorescent lights to obtain the proper spectrum and adjust the distance of these light sources from each cylinder to compensate for these potential problems.

10. Obtain three 100 mL cylinders with populations of *Daphnia* and place black paper sleeves around them as outlined earlier.

11. The colored filters can be placed directly over the opening of each cylinder, ensuring that only that portion of the visible light spectrum enters the cylinder.

12. For each of the three treatments, use a 150-watt incandescent bulb *and* two 25-watt fluorescent bulbs to obtain the proper spectrum. Set up each cylinder as follows:

Red: 3 red filters over a cylinder 30 cm away from the light source

Green: 1 green filter over a cylinder 25 cm away from the light source

Blue: 1 blue filter over a cylinder 10 cm away from the light source

13. After the treatments have been set up, turn on each light and allow the *Daphnia* in each cylinder 5 minutes to adjust their positions within the water columns.

14. After 5 minutes, remove the paper sleeve from each cylinder and quickly count the number of *Daphnia* in each 10 mL volume within the cylinder as before. Record your data in Table 11.7.

15. If time permits, repeat the experiment twice more, exposing each of the three tubes to a different wavelength (a different filter) each time to reduce the effect of any cylinder-specific variation in response.

TABLE 11.7 ▪ Number of *Daphnia* in Each Volume of the Water Column After 5 Minutes of Exposure to Different Wavelengths of Light

10 mL Volumes in Graduated Cylinder	Trial 1			Trial 2			Trial 3		
	Blue	Green	Red	Blue	Green	Red	Blue	Green	Red
90–100									
80–90									
70–80									
60–70									
50–60									
40–50									
30–40									
20–30									
10–20									
0–10									

Check Your Progress

1. Did the wavelength of light have any affect on the position within the water column that *Daphnia* selected?

2. Which wavelength caused *Daphnia* to select the highest position?

3. Based on these findings, would you retain or reject your hypothesis?

4. Propose additional studies that might be performed on *Daphnia* to delve further into these or other orientation behaviors.

Questions For Review

1. List at least one advantage of an exoskeleton.

2. What disadvantages associated with having exoskeletons did arthropods have to overcome to allow natural selection to favor this trait? How did they circumvent these problems?

3. Arthropods have a/an _____ circulatory system and a _____ digestive system.
   a. closed, complete
   b. open, complete
   c. closed, incomplete
   d. open, incomplete
   e. nonexistent, complete

4. List three major features that grasshoppers and crayfish share in common *and* three that are found in grasshoppers but not in crayfish.

   Grasshoppers and Crayfish: **Grasshoppers ONLY:**

   1.   1.

   2.   2.

   3.   3.

Echinodermata

After completing the exercises in this chapter, you should be able to:

1. Identify the internal and external anatomy of a sea star and discuss the functions of each structure.
2. Identify the internal and external anatomy of a sea urchin and discuss the functions of each structure.
3. Identify the internal and external anatomy of a sea cucumber and discuss the functions of each structure.
4. Characterize patterns of locomotion in sea stars and sea urchins.
5. Define all boldface terms.

The phylum Echinodermata includes sea stars, brittle stars, sand dollars, sea urchins, sea cucumbers and sea lilies. As their common names suggest, nearly all of the 6000 species comprising this phylum are marine organisms which possess many common features that represent adaptations to this primarily sessile lifestyle. Although their young are bilaterally-symmetrical, the adults in this phylum typically display a pattern of **radial symmetry** based on a five-point (pentamerous) design. Radial symmetry and the absence of cephalization have evolved in the echinoderm line as adaptations to the sedentary lifestyle these animals display. Other advanced features of echinoderms include: a **true coelom**, a hydraulic system for locomotion, the **water vascular system**, and an **endoskeleton** composed primarily of calcium carbonate **ossicles**.

For nearly 300 million years the echinoderms were among the most prevalent marine creatures in the oceans. An additional 13,000 species of echinoderms are known only from the fossil record as a testament to their former reign. Many groups, like the sea lilies, date back as far as 600 million years!

Though they do not enjoy the popularity they once had, living echinoderms have carved out a niche as slow-moving, bottom-dwellers with which few marine creatures can compete.

Despite their unusual appearance, echinoderms share a close evolutionary link to modern chordates. Both phyla are believed to have diverged from a common ancestor over 500 million years ago, and have since grown apart in appearance. Nonetheless, both share many unique characteristics, such as endoskeletons and bilaterally-symmetrical young. Another major feature that echinoderms and chordates have in common can be seen in their embryological pathways. Members of both groups are **deuterostomes**, meaning their mouths develop from the second embryonic opening and their embryonic cells divide by radial cleavage. In deuterostomes the anus develops from the first embryonic opening, the blastopore. Other invertebrates that you have examined (annelids, molluscs and arthropods) are **protostomes**, meaning that their mouths develop from the blastopore and their embryonic cells divide by spiral cleavage.

Sea Star Anatomy

Materials needed:

- preserved sea star (*Asterias*)
- dissecting tools
- dissecting pan
- dissecting microscope

Sea stars, often erroneously referred to as "starfish," belong to the class Asteroidea, a group of slow-moving, carnivorous predators that stalk their prey along rocky coastlines and coral reefs. The undersides of their arms are covered with tube feet which are used for both locomotion and prey capture. They possess a tenacious grip and can easily pry open the halves of a fully-closed clam or mussel to devour the tender meat inside! There are approximately 1,600 species of sea stars, making them the second largest group of echinoderms.

EXTERNAL ANATOMY

1. Obtain a preserved specimen of *Asterias* or another species of sea star and place it in your dissecting pan.

2. Your first task is to be able to distinguish the **oral** surface from the **aboral** surface. The "underside" of the animal containing the mouth and thousands of tube feet embedded in grooves along the arms is the oral surface. The aboral surface bears the madreporite—the off-center, external opening to the water vascular system (Table 12.1).

3. The anus is located in the center of the body, but is usually too small to be visible.

4. Close examination of the epidermis on the aboral surface should reveal small, pincer-like **pedicellariae** which the sea star uses to cleanse its skin surface and **dermal branchiae** which aid in gas exchange and excretion (by simple diffusion). A dissecting microscope may be necessary to view these minute structures.

5. Use Table 12.1 and Figure 12.1 to assist you in identifying the remaining external features of the sea star.

TABLE 12.1 ▪ External Anatomy of the Sea Star

Structure	Function
Spines	Calcareous projections for protection and support
Pedicellariae	Pincer-like structures believed to kill small organisms that might settle on body surfaces, thus keeping the epidermis free of parasites and algae
Dermal branchiae	Gas exchange and excretion through simple diffusion
Anus	Regulates egestion of undigested food (feces) from the body
Madreporite	Porous entrance to the water vascular system that serves as both pressure regulator and simple filter
Ambulacral grooves	Radiate from the mouth to the tip of each arm and house the tube feet
Tube feet	Locomotion and prey capture
Mouth	External opening to cardiac chamber (through a short esophagus)

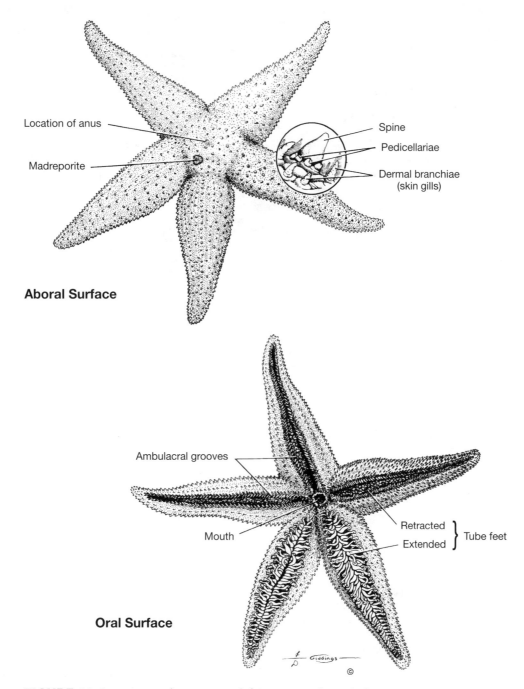

Location of anus

Madreporite

Aboral Surface

Spine

Pedicellariae

Dermal branchiae
(skin gills)

Ambulacral grooves

Mouth

Retracted
}
Extended   } Tube feet

Oral Surface

Giddings
©

FIGURE 12.1 ▪ External anatomy of the sea star (*Asterias*).

INTERNAL ANATOMY

1. Use a sharp pair of dissecting scissors to cut away the epidermis from the *aboral* side of one of the arms and the central disk of your sea star.

2. ***Do not*** rip away the epidermis; rather gently tease it apart from the underlying tissue with a dissecting needle or blunt probe, especially in the region of the central disk (Fig. 12.2a). The underlying organs lie very close to the inner surface of the epidermis and actually adhere to it in some places. Sloppy technique here may ruin underlying organs!

3. Next select a different arm and carefully cut it off near its base with your scissors to reveal a cross-sectional view of the arm's interior (Fig. 12.2c).

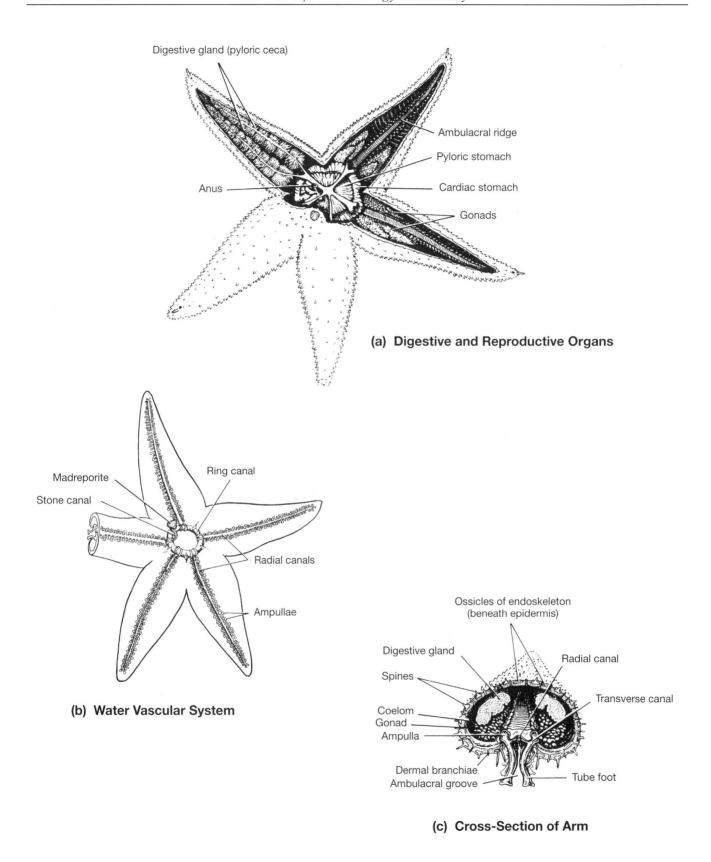

Digestive gland (pyloric ceca)

Ambulacral ridge

Pyloric stomach

Cardiac stomach

Anus

Gonads

(a) Digestive and Reproductive Organs

Madreporite

Stone canal

Ring canal

Radial canals

Ampullae

(b) Water Vascular System

Ossicles of endoskeleton
(beneath epidermis)

Digestive gland

Radial canal

Spines

Transverse canal

Coelom
Gonad
Ampulla

Dermal branchiae
Ambulacral groove

Tube foot

(c) Cross-Section of Arm

FIGURE 12.2 ▪ Aboral view of the internal anatomy of a sea star depicting (a) digestive and reproductive organs, (b) water vascular system and (c) cross-sectional view of arm.

4. The internal anatomy of each arm is identical, so careful examination of only one arm is necessary. Work from the aboral surface of the arm toward the oral surface, identifying each structure or organ as you proceed (Table 12.2).

A challenge faced by sea stars in their developmental transformation from a bilateral juvenile into a radial adult is the reorganization and dorso-ventral compression of their digestive system. Due to their compressed body shape and the limited distance between the mouth and anus, the coelomic spaces in the arms have been recruited to house the bulk of the digestive system. The digestive system of the sea star is comprised of the following organs: **mouth, digestive glands, pyloric stomach, cardiac stomach** and **anus.** The large digestive glands located in the arms are connected to the pyloric stomach through thin, membranous connections which may still be intact in your specimen (Fig. 12.2a). The reproductive system of the sea star consists of pairs of **gonads** housed along the interior, oral surfaces of each arm. Sea stars, like all echinoderms are **dioecious,** but the sexes are difficult to differentiate through dissection. Fertilization is external; thus males and females must coordinate the timing of their reproductive efforts to ensure fertilization of the eggs.

5. To view the water vascular system, you must remove the components of the digestive and reproductive systems from the central disk and arm.

The **water vascular system** of the sea star is fairly representative of echinoderms in general. It consists of a series of internal canals which branch from a centralized **ring canal** into primary **radial canals** that terminate into the hundreds of **tube feet** that line the **ambulacral grooves** along the oral surfaces of the arms. Sea water enters this system through the **madreporite** and is kept under pressure by muscular contractions by the sea star to effect an entire repertoire of behaviors, including: locomotion, reproduction, defense and prey capture (Fig. 12.2b).

TABLE 12.2 ▪ Internal Anatomy of the Sea Star

Structure	Function
Digestive glands (pyloric ceca)	Secrete digestive enzymes for breakdown of food; play a major role in absorption and storage of food materials
Pyloric stomach	Receives secretions of digestive glands for chemical digestion
Cardiac stomach	Can be everted through the mouth to envelope prey; site of initiation of digestion
Gonads	Produce gametes for reproduction
Anus	Regulates egestion of undigested food (feces) from the body
Ring canal	Portion of water vascular system encircling the mouth
Radial canals	Portions of water vascular system emanating from the ring canal and leading into each arm of the sea star
Stone canal	Portion of water vascular system leading from the madreporite to the ring canal
Madreporite	Porous entrance to the water vascular system that serves as both pressure regulator and simple filter
Transverse canals	Portions of water vascular system connecting radial canals with pairs of tube feet
Tube feet	Locomotion and prey capture
Ambulacral grooves	Radiate from the mouth to the tip of each arm and contain the tube feet
Dermal branchiae	Gas exchange and excretion
Ampullae	Provide hydraulic pressure for movement of the tube feet
Ossicles of endoskeleton	Support

Check Your Progress

1. What role does the madreporite play in the water vascular system of sea stars?

2. Does the sea star possess a heart or circulatory system? How do you suppose nutrients and oxygen are distributed to the tissues of the body?

3. Does the sea star possess an excretory system? How do you suppose metabolic wastes are eliminated from the body? (And don't say "through the anus.")

4. Why are ossicles classified as an endoskeleton?

EXERCISE 12–2   **Photographic Atlas Reference Page 81**

Sea Urchin Anatomy

> *Materials needed:*
> - preserved sea urchin
> - dried sea urchin tests
> - dissecting tools
> - dissecting pan

Sea urchins are members of the class Echinoidea, which also includes sand dollars and the lesser known heart urchins, totaling less than 1,000 species in all. The most distinguishing feature of sea urchins is their long, rigid, moveable **spines** used primarily in locomotion and defense. In most species, the **ossicles** underneath the epidermis are fused together forming large plates which constitute the solid, inflexible **test** that also characterizes urchins. As an urchin grows, its test enlarges in all directions along sutures, much the way a child's skull grows inside its scalp.

EXTERNAL ANATOMY

1. Obtain a preserved sea urchin and examine its external features using Table 12.3 as a guide.

A **mouth** bearing prominent teeth used to rasp algae from the substrate is located on the oral side. Surrounding the mouth is a thin, fleshy membrane known as the **peristome** which everts and retracts the teeth (Fig. 12.3). **Tube feet** occur in rows along the spheroid body and **pedicellariae** are present for gleaning the body surface of parasites and foreign debris. Respiration occurs by means of gills located in rows along the body. The **spines** are attached at their bases by "ball-and-socket" joints to small muscles which coordinate locomotion with the extensible tube feet. In some species the spines exude toxins and serve as both a chemical and physical defense system for the urchin, allowing them to wedge securely into crevices between rocks.

TABLE 12.3 ▪ External Anatomy of the Sea Urchin

Structure	Function
Mouth	External opening to oral cavity
Teeth	Scrape (or rasp) food into mouth
Gills	Gas exchange
Peristome	Circular oral membrane which allows for eversion of teeth for feeding
Spines	Defense and locomotion
Tube feet	Locomotion
Pedicellariae	Pincer-like structures which glean the body surface and may contain poison glands for defense
Test	Support structure composed of numerous calcareous plates located beneath the epidermis and constituting an endoskeleton

2. If a dried sea urchin test is available, locate the oral and aboral surfaces. Notice that surrounding the anus are five small **genital pores** (gonopores) through which gametes are released. You may also be able to see the **madreporite**, in urchins a very small opening through which water enters the water vascular system.

INTERNAL ANATOMY

1. Using a pair of sharp dissecting scissors, insert the pointed end into the peristome adjacent to the mouth and make a longitudinal incision from the peristome to the center of the aboral side (where the anus is located).

2. Make another incision parallel to the first about 1/3 of the way around the body, so that you are essentially removing one third of the test and exposing the internal anatomy.

The well-developed **coelomic cavity** is quite evident and is filled with little other than a simple digestive system and reproductive organs (Fig. 12.4). Digestion in urchins begins as their **teeth** rasp food into the mouth off rocks. Internally, the movements of the mouth and teeth are controlled by a complex system of ossicles and muscles known as **Aristotle's lantern**. Food passes along an **esophagus** to an enlarged portion of the

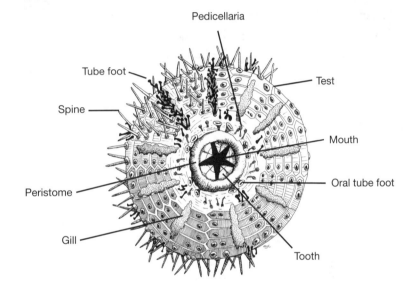

FIGURE 12.3 ▪ Oral view of the external anatomy of a sea urchin.

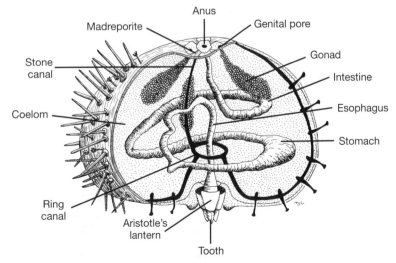

FIGURE 12.4 ▪ Lateral view of the internal anatomy of a sea urchin.

intestine which may, for convenience, be referred to as a **stomach** (Table 12.4). Urchins lack the complex stomach system of sea stars; rather the esophagus leads into a very long, convoluted **intestine** where the bulk of digestion and nutrient absorption occurs. Indigestible items pass through the **anus** as they are released from the body.

Gonads should be visible as yellowish, granular masses attached to the inside of the aboral surface.

Sea urchins are **dioecious** (as you may remember from the fertilization exercises in Chapter 3) and sperm or eggs are released through the genital pores on the aboral surface. Fertilization is external. The water vascular system of urchins consists of the **madreporite, stone canal, ring canal** and **lateral canals,** the latter being fused to the inner surface of the test along rows that correspond to the arrangement of tube feet protruding through the body wall.

TABLE 12.4 ▪ Internal Anatomy of the Sea Urchin

Structure	Function
Genital pores	Small openings on the aboral surface that release gametes
Gonads	Produce gametes for reproduction
Intestine	Final digestion of food received from stomach and subsequent nutrient absorption
Esophagus	Transports food from mouth to stomach
Stomach	Enlarged portion of intestine for food storage and initiation of digestion
Teeth	Scrape (or rasp) food into mouth
Aristotle's lantern	Anchors teeth and coordinates their movement for scraping and chewing
Ring canal	Portion of water vascular system encircling the esophagus
Stone canal	Portion of water vascular system leading from the madreporite to the ring canal
Madreporite	Porous entrance to the water vascular system that serves as both pressure regulator and simple filter
Anus	Regulates egestion of undigested food (feces) from the body

Check Your Progress

1. What is Aristotle's lantern?

2. Given that sea urchins lack arms, how do they move?

3. What differences in the digestive system of sea urchins (from that of carnivorous sea stars) reflect their herbivorous lifestyle?

Sea Cucumber Anatomy

> *Materials needed:*
> - preserved sea cucumber
> - dissecting tools
> - dissecting pan

Sea cucumbers belong to the class Holothuroidea. This class contains just over 1,000 species of soft-bodied, filter-feeding echinoderms. Sea cucumbers superficially resemble elongated urchins with their spines removed and their ossicles reduced to microscopic structures embedded deep within the body wall.

1. Obtain a preserved sea cucumber and place it in your dissecting pan.

2. Notice the warty, leathery texture of its skin. Unlike most echinoderms, sea cucumbers have a soft body and are generally bilaterally symmetrical with definite cranial (anterior) and caudal (posterior) ends (Fig. 12.5).

Tube feet occur sparsely along the body wall and are organized into longitudinal rows. Near the mouth, the tube feet are modified into feathery **tentacles** which sea cucumbers use to capture food particles from the surrounding water. Sea cucumbers are sometimes categorized as deposit-feeders, since they are able to extract organic matter from the fine mud ooze of the deep sea floor.

3. Using dissecting scissors, make two parallel, longitudinal incisions along the body wall from the tentacles toward the anus, separating the body into two equal halves.

4. Use Table 12.5 and Figure 12.5 to assist you in identifying the internal anatomy of your sea cucumber.

The digestive system of sea cucumbers is similar to that of urchins, containing a muscularized **pharynx** (or esophagus) and a greatly elongated **intestine** leading to an **anus**. The water vascular system follows the typical echinoderm pattern, except for the **madreporite** which lacks a direct connection to the outside of the body. Instead it floats freely in the large coelomic cavity; thus the

water vascular system of sea cucumbers uses coelomic fluid rather than sea water as its pressure generating medium. Sea cucumbers, like most echinoderms, lack a circulatory system and use coelomic fluid as the primary circulatory medium as well. A single, filamentous **gonad** is present near the cranial end of the body and gametes are released through a gonopore located near the base of the tentacles. Paired unique structures, found only in holothurians, are the **respiratory trees**—highly-branched outgrowths of the hindgut. Typically two of these structures are present and are connected to the cloaca. Water is pumped into and out of the respiratory trees by muscular contractions of the anus, and gas exchange occurs across the thin tissues of the respiratory trees as water passes over them.

No discussion of sea cucumbers would be complete without mentioning their amazing ability to

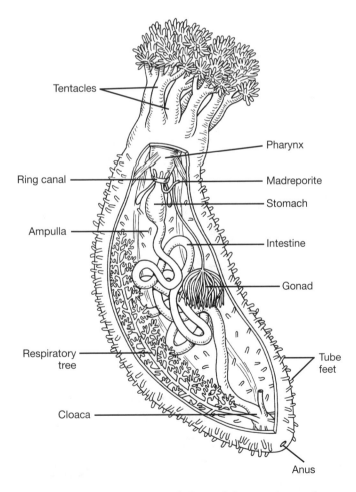

FIGURE 12.5 • Lateral view of the anatomy of a sea cucumber.

eviscerate their internal organs. When sufficiently disturbed or provoked, the muscular body wall quickly contracts and some portion of the viscera is forced out through the anus, essentially turning the sea cucumber inside out! In some species this is limited to the expulsion of the respiratory trees, but in others, true evisceration occurs and the entire contents of the body cavity are expelled. Eventually the lost body parts are reformed, reflecting the astounding regenerative capabilities of echinoderms!

TABLE 12.5 ▪ Anatomy of the Sea Cucumber

Structure	Function
Tentacles	Food collection
Ring canal	Portion of water vascular system encircling the junction of pharynx and stomach
Ampullae	Provide hydraulic pressure for movement of the tube feet
Pharynx	Transports food from mouth to stomach
Madreporite	Entrance to the water vascular system that serves as a pressure regulator; floats freely within the coelomic cavity and is not connected to the outer body wall
Stomach	Extremely reduced in holothurians; site of food storage and initiation of digestion
Intestine	Final digestion of food received from stomach and subsequent nutrient absorption
Gonad	Produces gametes for reproduction
Tube feet	Locomotion
Respiratory trees	Gas exchange
Cloaca	Pumps a ventilating current of seawater into and out of respiratory trees
Anus	Regulates egestion of undigested food (feces) from the body

Check Your Progress

1. What method of food acquisition do sea cucumbers employ?

2. Since sea cucumbers have lost the sharp spines and hard endoskeleton on which most echinoderms rely for protection, how do they defend themselves?

3. What characteristics do sea cucumbers possess that would cause biologists to group them in the same phylum with other echinoderms?

Locomotion in Sea Stars and Sea Urchins

If living sea stars and sea urchins are available in your laboratory, you can perform some simple exercises to characterize the different methods of locomotion these two groups use. Despite the simple, decentralized nature of the echinoderm nervous system, they are clearly capable of sophisticated, coordinated behaviors and movements. There is even evidence to support the contention that a limited degree of learning is possible in some echinoderms!

1. First obtain a living sea urchin and place it in a large finger bowl filled with sea water.

2. Observe its pattern of locomotion and answer the following questions.

   Are the spines utilized in locomotion?

   Are the tube feet utilized in locomotion?

3. Now, gently roll the sea urchin over on its "side." Be careful not to pull too hard on it as you dislodge the urchin from the glass container or you will rip off its tube feet!

   Can the sea urchin move as effectively on its "side" as in its former orientation or does it display a tendency to right itself to its former position?

   Do you suppose that sea urchins can move fairly easily in any orientation?

4. Now obtain a living sea star and place it in another large finger bowl containing sea water.

5. Observe its pattern of locomotion and answer the following questions.

   Are the spines utilized in locomotion?

   Are the tube feet utilized in locomotion? Do the tube feet move in random fashion or in smooth, coordinated waves?

Do the arms move appreciably during locomotion?

6. Now gently flip the sea star over onto its aboral surface, again being careful to gently dislodge it from the container.

   Can the sea star move as effectively on its aboral side as in its former orientation?

   Did the sea star "right itself" to its former position with the oral surface facing downward?

   Do you suppose that sea stars can move with equal ease in any orientation?

7. Once again, gently flip the sea star over onto its aboral side. This time, as it rights itself, pay close attention to which of its arms it uses in the righting response.

8. Repeat this flipping process several more times—each time paying careful attention to which arm (or arms) the sea star uses to accomplish the righting response. You may find it helpful to place a small identifying mark on the arm or arms that you wish to watch, especially if you can not distinguish them from the other arms.

   Does the sea star favor one arm over the others when righting itself?

   What does this tell you about a sea star's ability to sense directional stimuli such as gravity?

   What does this tell you about a sea star's ability to distinguish among its arms and coordinate their movements?

   How is this similar to dominant "handedness" in humans?

Questions For Review

1. Echinodermata is the first phylum you have examined whose members possess true endoskeletons. What are some advantages that an endoskeleton provides?

2. What is the primary function of the water vascular system?

3. What is unique about the madreporite of sea cucumbers (Holothuroidea)?

4. Fill in Table 12.6 with the appropriate characteristics for each class of echinoderms covered to compare the adaptations to different lifestyles present among members of these groups.

TABLE 12.6 ▪ Comparison of Major Characteristics Among Echinoderm Classes

	Class		
	Asteroidea	**Echinoidea**	**Holothuroidea**
Shape of arms			
Development of tube feet			
Development of ossicles			
Feeding method			
Spine structure			

Chordata

After completing the exercises in this chapter, you should be able to:

1. Identify the major external features of the lancelet and compare them to homologous structures in other chordates.

2. Identify the major internal organs in the lancelet, discuss their functions in the body and relate them to homologous organs in other chordates.

3. Define all boldface terms.

The phylum Chordata is a large, familiar phylum containing a tremendously diverse group of nearly 45,000 species. We are chordates, as are birds, dogs, lizards, frogs, fish, sharks, turtles and elephants. Thus the remainder of this book has been devoted to exploring the intricate differences among the major classes of chordates. Having said that, it should be pointed out that all chordates do possess several traits in common. At some point in their lifecycles, all chordates have (1) a notochord, (2) a dorsal, hollow nerve cord, (3) pharyngeal gill slits and (4) a postanal tail. Since all chordates share a recent common evolutionary ancestor, many of the morphological features of the members of these groups are homologous. **Homologous structures** are structures in different species that are similar due to common ancestry. The four chordate characteristics listed above are all examples of homologous structures. As you proceed through the next several chapters, keep this in mind and make an attempt to identify the many homologies seen in each group.

In this chapter you will examine a common member of the subphylum Cephalochordata, the lancelet (*Branchiostoma* or *Amphioxus*). Cephalochordates display several intermediate characteristics between invertebrates and vertebrates, making them a useful organism for understanding the evolutionary transition that occurred in the ancestral line of chordates. All four chordate characteristics are present in the adult stage, yet lancelets lack the vertebral column or cranium of vertebrates. This small subphylum contains around two dozen species of small, laterally-flattened, fishlike filter-feeders that are commonly found in shallow marine and brackish waters. They spend the majority of their time partially buried in the sand with only their heads protruding above the sediment. They do possess the ability to swim by what could best be described as an erratic, convulsive twitching motion. Nonetheless, their ability to swim is tantamount to dispersal and mating.

Cephalochordate Anatomy

> *Materials needed:*
> - prepared slide of lancelet w.m.
> - prepared slide of lancelet c.s.
> - compound microscope

EXTERNAL ANATOMY

1. Obtain a prepared slide of a whole mount of a lancelet and view it using low power on your compound microscope.

2. Start identifying the structures at the cranial end of your specimen and progress caudally using Table 13.1 and Figures 13.1 and 13.2 to assist you.

Locomotion in this group is achieved by alternating muscular contractions of the segmented **myomeres** against the fairly rigid **notochord**. As myomeres along one side of the body contract, opposing myomeres along the other side of the body relax and the body bends. These lateral undulations are enhanced by the presence of small dorsal and ventral fins which aid in steering and navigation. The same motions employed for swimming also double as the lancelet's mechanism for burrowing into the soft sand for protection and feeding. During feeding, a current of water is driven into the **mouth** through cilia on the **oral cirri** and **wheel organ**, bringing with it suspended food particles which are filtered across the **gill bars**. As small food particles are trapped by the mucus coating on the gill bars, the water passes through the **gill slits** into the **atrium** and flows caudally along the body, eventually exiting

TABLE 13.1 ▪ Lateral View of Lancelet Anatomy

Structure	Function
Oral hood	Surrounds mouth and supports cirri for intake of water and minute food particles
Rostrum	Cranial projection shielding entrance to the mouth dorsally
Wheel organ	Lined with cilia which produce a current of water that brings food into the mouth
Oral cirri	Act as a strainer to exclude larger particles from the mouth (filter feeding)
Mouth	External opening to the pharynx
Gill bars	Have thin mucus coating which traps ingested food particles; also contain blood vessels supplementing gas exchange for respiration
Gill slits	Allow for outflow of water from pharynx between gill bars for food capture and respiration
Velum	Transverse partition encircling mouth
Pharynx	Receives incoming water from mouth, entraps food in a film of mucus and is responsible for some gas exchange
Hepatic cecum	Lateral outpocket of intestine responsible for intracellular digestion of small food particles and lipid and glycogen storage
Intestine	Site of enzymatic digestion of larger food particles
Atrium	Ventral body chamber that receives water passed from pharynx through gill slits
Atriopore	Discharges water from the atrium to the external environment
Anus	Regulates egestion of undigested food (feces) from the body
Ventral fin	Steering while swimming and burrowing
Caudal fin	Steering while swimming and burrowing
Dorsal fin	Steering while swimming and burrowing
Notochord	Provides rigid support for anchoring myomeres to provide lateral body movements
Myomeres	Provide muscular movements for swimming and burrowing
Dorsal nerve cord	Handles majority of nervous coordination without intervention of the brain

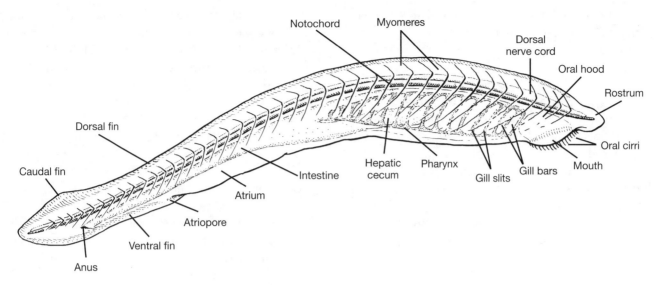

FIGURE 13.1 ▪ Lateral view of the anatomy of the lancelet (*Branchiostoma* or *Amphioxus*).

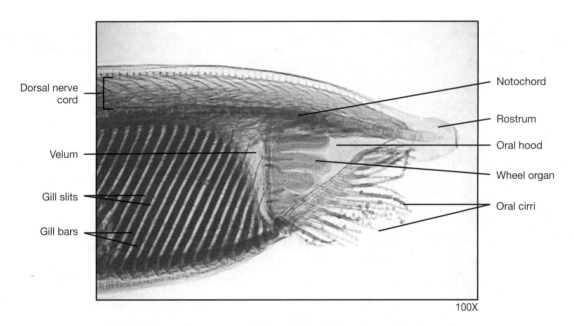

FIGURE 13.2 ▪ Lateral view of cranial region of the lancelet.

the body through the ventral **atriopore**. Trapped food particles, on the other hand, are transported along the dorsal aspect of the **pharynx** into the **intestine** for digestion and nutrient absorption. Digestion is initially extracellular, but is completed intracellularly in the walls of the intestine and the small, lateral outpocket of the intestine, known as the hepatic cecum. In addition to digestion, the **hepatic cecum** functions in lipid and glycogen storage and is believed to be the evolutionary precursor of the vertebrate liver. Indigestible food particles pass through the intestine and exit the body through the **anus**.

INTERNAL ANATOMY

1. Obtain a prepared slide of a cross-section of the lancelet and view it using low power on your compound microscope.

2. Use Table 13.2 and Figure 13.3 to help you identify the internal structures visible on the cross-section.

Like all chordates, lancelets possess a **closed circulatory system** consisting of a series of closed vessels for blood flow. Unlike most animals with closed circulatory systems, there is no heart. Occasionally the paired **dorsal aortae** or ventral aortae are visible on cross-sections through the body, but a full compliment of arteries and veins exists. The blood contains no respiratory pigments and is thus colorless, functioning primarily in nutrient distribution, rather than in gas exchange and transport. The **gills** play a minor role in respiration, but the bulk of gas exchange occurs across thin flaps of highly vascularized tissue along the ventral surface of the body wall. Excretion is accomplished through segmented **nephridia** akin to those seen in annelids. Metabolic waste is collected in the nephridia and channeled to the **atriopore** for elimination from the body. In lancelets the sexes are separate, but structurally very similar. They possess from 25 to 38 pairs of serially arranged **gonads** along the ventral region of the body wall in the atrium, lateral to the pharynx. A hallmark of chordates is the presence of a hollow **nerve cord** running along the dorsal surface of the body. Aside from this feature, lancelets have little else in the way of a central nervous system. Cephalization is absent; instead, segmentally arranged nerves handle the majority of muscular coordination along the body axis.

TABLE 13.2 ▪ Cross-sectional View of Lancelet Anatomy

Structure	Function
Epidermis	Outermost layer of cells providing protection and support
Notochord	Provides rigid support for anchoring myomeres to provide lateral body movements
Hepatic cecum	Lateral outpocket of intestine responsible for intracellular digestion of small food particles and lipid and glycogen storage
Pharynx	Receives incoming water from mouth, entraps food in a film of mucus and is responsible for some gas exchange
Gill slits	Allow for outflow of water from pharynx between gill bars for food capture and respiration
Dorsal nerve cord	Handles majority of nervous coordination without intervention of the brain
Myomeres	Provide muscular movements for swimming and burrowing
Dorsal aortae	Paired blood vessels which join behind the pharynx to form the median dorsal aorta which carries nutrient-laden blood caudally to the body tissues and intestine
Nephridia	Ciliated ducts which transport metabolic wastes from the dorsal portions of the coelom to the atrium
Gill bars	Have thin mucus coating which traps ingested food particles; also contain blood vessels supplementing gas exchange for respiration
Atrium	Ventral body chamber that receives water passed from pharynx through gill slits
Testes (male)	Produce sperm for external fertilization
Ovaries (female)	Produce eggs for external fertilization

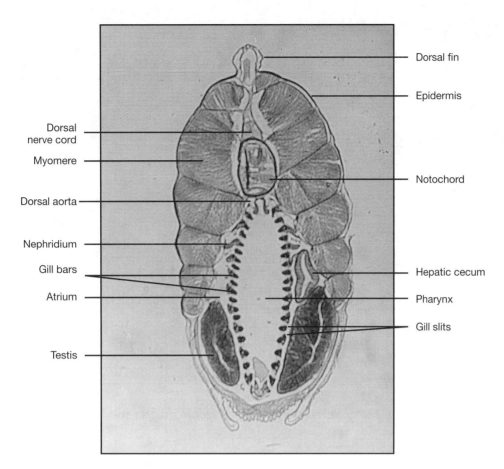

FIGURE 13.3 ▪ Cross-section through pharyngeal region of the lancelet.

Questions For Review

1. List the 4 characteristics common to all chordates.

    a.

    b.

    c.

    d.

2. Which structures in the lancelet exhibit segmentation?

3. What other animals have you encountered that use filter feeding to get food?

4. What aspects of the circulatory system of lancelets have been modified in "higher" chordate groups?

5. How does gas exchange occur in the lancelet?

6. In lancelets, water that has filtered across the gill bars and has picked up metabolic waste products from the nephridia exits the body through the _____, while indigestible food is released from the body through the _____.

14

Osteichthyes

After completing the exercises in this chapter, you should be able to:

1. Identify the major external features of a bony fish and compare them to homologous structures in other vertebrates.

2. Identify the major skeletal features of a bony fish and compare them to homologous bones in other vertebrates.

3. Identify the major internal organs in a bony fish, discuss their functions in the body and relate them to homologous organs in other vertebrates.

4. Define all boldface terms.

All fish belong to the subphylum Vertebrata and exemplify organisms that are highly adapted to aquatic environments. Jawless fishes, such as lampreys and hagfish, are the most primitive of living fish and belong to the superclass Agnatha, while sharks, rays and skates are cartilaginous fish in the class Chondrichthyes. This chapter focuses on bony fish within the class Osteichthyes. This single class of vertebrates contains nearly 25,000 living species—over half of all known vertebrates. That's more than all jawless fish, cartilaginous fish, amphibian, reptile, bird and mammal species combined! The class Osteichthyes includes everything from freshwater species such as bass, trout, perch and minnows to marine species such as flounder, grouper, tarpon, sailfish, and even sea horses!

Modern bony fish possess many of the familiar vertebrate characteristics such as: a bilaterally-symmetrical body, a well-developed head bearing sensory organs and a terminal mouth, a skull enclosing the brain, a vertebral column enclosing the dorsal nerve cord and a closed circulatory system with a chambered heart. Bony fish do, however, have an array of characteristics that represent extreme specialization to the aquatic world in which they have been evolving for hundreds of millions of years. Many of these features are homologies shared with the other vertebrate groups that have been modified in fishes to best equip them to cope with the challenges they face under water.

EXERCISE 14–1 **Photographic Atlas Reference Pages 98–101**

Anatomy of a Bony Fish

Materials needed:
- mounted bony fish skeleton
- preserved perch
- dissecting tools
- dissecting pan
- compound microscope
- clean glass slides
- coverslips

EXTERNAL ANATOMY

1. Obtain a preserved bony fish such as a perch and examine its external features. Perch are common inhabitants of freshwater lakes, ponds and occasionally streams and are close relatives of bass, crappie and other sunfish.

2. Use Table 14.1 and Figure 14.1 to assist you in identifying the external anatomy of your fish.

3. As you examine your specimen, try to draw comparisons between the many structures which are shared by other familiar vertebrates.

Notice that fish have a streamlined, **fusiform body** shape designed to minimize drag during swimming. The body is thickest about one-third of the way back from the head and tapers in both directions. The large head bears two **nostrils**, two large **eyes** that lack eyelids, a terminal **mouth** equipped with small **teeth** and bony **opercula** covering the feathery **gills** on each side of the head. Locate the **anus** on the ventral surface of the body cranial to the anal fin. In addition to an anus, female perch possess a single **urogenital opening** for the release of eggs and metabolic wastes. Males have a separate genital pore and urinary opening.

4. Remove a **scale** from your specimen and view a wet mount of the scale using low power. Notice the numerous concentric ridges on the scale. Fish scales grow continuously throughout the fish's life and these ridges actually represent growth rings. As the fish ages, new material is deposited around the margin of each scale forming the pattern you see. For this reason, biologists can determine the age of a fish by counting these growth rings.

TABLE 14.1 ▪ External Anatomy of the Bony Fish

Structure	Function
Eyes	Large image-forming sight organs, lacking eyelids
Nostrils	Paired openings in dorsum of head leading to olfactory receptors
Mandible	Lower jawbone bearing teeth for prey capture
Maxilla	Upper jawbone, fused to skull, bearing teeth for prey capture
Opercula	Paired bony plates which cover the gills on either side of the head, attached cranially and dorsally but open caudally and ventrally for the release of water
Pectoral fins	Steering and braking while swimming and maintenance of dorsal-ventral orientation while suspended
Pelvic fins	Steering while swimming
Anal fin	Steering while swimming
Caudal fin	Provides thrust and acts as a rudder while swimming
Adipose fin	Vestigial remnant of second dorsal fin
Dorsal fin	Steering and maintenance of dorsal-ventral orientation while swimming
Lateral lines	Specialized sensory organs which detect vibrations and current directions in the water

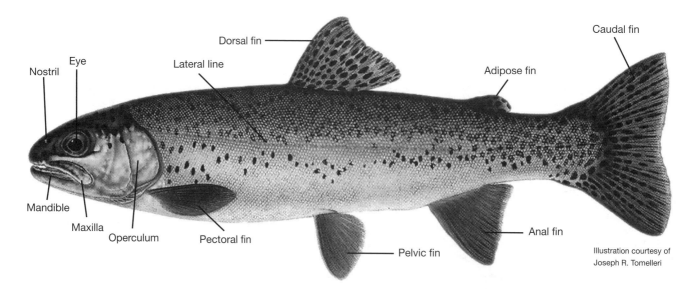

Illustration courtesy of Joseph R. Tomelleri

FIGURE 14.1 ▪ External anatomy of a bony fish (*Salmo gairdneri*, rainbow trout).

Check Your Progress

1. Is your specimen male or female?

2. Does most of a fish's power for movement come from the tail or from other fins?

3. What features of the head represent homologies to other vertebrate groups?

4. Which fins would be used to slow the fish after a quick burst of speed?

5. What is the function of the lateral lines along the fish's body?

Photographic Atlas Reference Page 100

SKELETAL SYSTEM

1. Obtain a mounted, articulated skeleton of a bony fish.
2. Examine it carefully using Figure 14.2 to assist you in identifying the skeletal elements present.

Notice that the skull is very heavily ossified and consists of both a **dermal exoskeleton** and a **bony endoskeleton** to provide the rigid, protective encasement for the brain and sensory organs. The jaws of carnivorous bony fish, such as perch, have extensive hinging to allow them to open widely enough to accommodate large prey. Some fish are capable of swallowing prey nearly as large as their own bodies! Food items are not chewed in the mouth, but swallowed whole by perch; thus teeth are generally small, numerous and sharp and function mainly in preventing prey from escaping as they are swallowed.

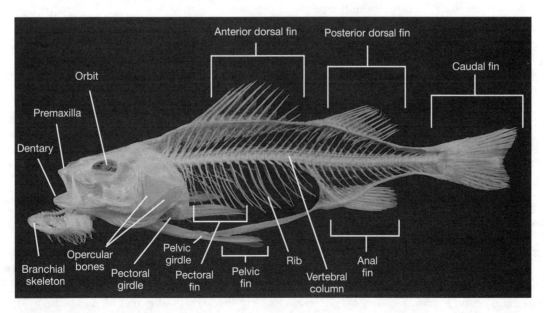

FIGURE 14.2 ▪ Skeletal system of a bony fish (perch).

Check Your Progress

1. What evidence of segmentation do you see in the perch skeleton?

2. Are all of the fins supported by bony elements?

3. Does the vertebral column extend all the way to the caudal fin?

4. Where is the skeleton reduced in its degree of coverage?

Photographic Atlas Reference Page 101

INTERNAL ANATOMY

1. Carefully cut away the bony operculum from one side of the perch to expose the **gills**. Four gills are present beneath the operculum and each gill consists of numerous **gill filaments** that extend caudally from the gill arch. These filaments are highly vascularized and provide a large surface area for contact with the water to facilitate gas exchange.

2. Remove one of the gills and examine it closely. Notice the hard, serrated structures projecting cranially. These structures are the **gill rakers** which protect the gill apparatus and prevent the passage of coarse material across the delicate gill filaments.

3. Next use a scalpel to make a shallow, longitudinal incision along the ventral midline of the body starting just cranial to the anus. Extend your incision through the pelvic girdle to the gills.

4. Then make a second, perpendicular incision from the origin of the first incision dorsally across the body to the level of the lateral line.

5. Make a third incision, similar to the last, from the midventral line just behind the gills across the body to the level of the lateral line.

6. Raise this "flap" of body wall and identify the internal organs listed in Table 14.2 and depicted in Figure 14.3. You may need to cut through the peritoneum to see the organs inside.

7. The large, whitish **liver** will probably be the most prominent organ in this region. After identifying the liver, reflect it dorsally (or remove it) to view the underlying organs.

8. The heart is ventral to the gills and cranial to the liver. You may need to extend your midventral incision further cranially. To access the perch's **two-chambered heart**, you will need to open the pericardial cavity.

The heart is composed of a larger, thin-walled **atrium** and a smaller, thick-walled **ventricle**. Deoxygenated blood returns from the body tissues, collects in the atrium, is passed into the ventricle and is forced out of the heart by contractions of the ventricle through the **ventral aorta** to the gills. Fish blood contains the respiratory pigment hemoglobin (found

TABLE 14.2 ▪ Internal Anatomy of the Bony Fish

Structure	Function
Tongue	Manipulation of food as well as chemosensory reception
Gills	Contain capillary beds for gas exchange for respiration
Heart	Two-chambered organ (one atrium, one ventricle) that pumps deoxygenated blood to the gills for oxygenation and from there throughout the various organ systems of the body
Liver	Large whitish organ that detoxifies many constituents of the absorbed digested compounds and functions in lipid and glycogen storage
Spleen	Elongate organ that stores blood and recycles worn-out red blood cells
Pyloric ceca	Three short pouches extending laterally from the small intestine near its juncture with the stomach that increase digestive surface area of the intestine
Stomach	Site of food storage and initiation of digestion
Intestine	Final digestion of food received from stomach
Testes (male)	Paired organs which produce sperm for transport through the paired vasa deferentia and release through the genital pore for external fertilization
Ovary (female)	Single (fused) organ which produces eggs for transport through the short oviduct and release through the urogenital pore for external fertilization
Anus	Regulates egestion of undigested food (feces) from the body
Urinary bladder	Storage organ for ammonia prior to elimination through urogenital opening
Hypaxial muscle	Muscle segments ventral to the transverse septum; used for swimming
Epaxial muscle	Muscle segments dorsal to the transverse septum; used for swimming
Vertebrae	Columnar units of the vertebral column which provide support, house the dorsal nerve cord and articulate with the ribs
Swim bladder	Hollow, gas-filled sac that serves as a buoyancy organ
Spinal cord	Part of the central nervous system which leads from the brain to the tail and conducts nerve impulses between the brain and the peripheral nervous system
Ureters	Paired tubes which transport kidney filtrate (ammonia) to the urinary bladder
Kidneys	Paired organs which filter nitrogenous wastes from the blood

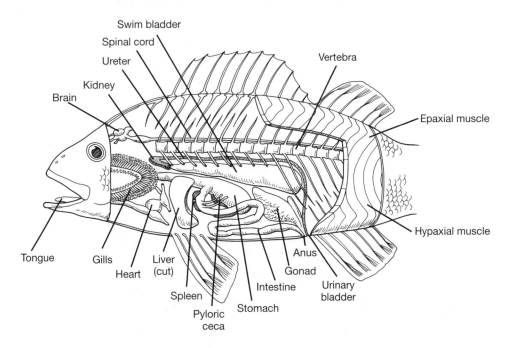

FIGURE 14.3 ▪ Internal anatomy of a bony fish (perch).

in all vertebrates) which increases the oxygen-carrying capacity of the blood and gives the blood its familiar red color. Oxygenation of the blood occurs in the gills and the blood then travels throughout the rest of the body before returning to the heart.

9. To examine the **brain** of the perch, you must first remove the skin from the dorsal surface of the skull.

10. Next, use a scalpel to carefully shave away the bony roof of the skull above the brain, flake by flake, until you have penetrated the brain case. This process is quite tedious and will take some time to complete.

11. The brain is encased in an outer gelatinous covering and an inner pigmented membrane—both of which must also be removed before you can

view the five regions of the brain. These regions are listed in Table 14.3 in order from most rostral to most caudal.

TABLE 14.3 ▪ Regions of the Perch Brain

Structure	Function
Telencephalon	Small region containing olfactory lobes and cerebral hemispheres
Diencephalon	Hypothalamus, thalamus, pineal body and pituitary gland
Mesencephalon	Largest region of the brain; comprised of optic lobes
Metencephalon	Second largest region of the brain; comprised of cerebellum
Myelencephalon	Medium-sized, posterior region comprised of medulla oblongata

 Questions For Review

1. Opercula are absent in cartilaginous and jawless fishes and represent an evolutionary advancement seen in bony fish. How do opercula allow a bony fish to pass water over its gills while the fish remains stationary? Why would this be viewed as an adaptation?

2. The swim bladder is another evolutionary advancement found in bony fish. What advantage do fish with swim bladders have?

3. What anatomical evidence supports the contention that fish have keen senses of vision and smell?

4. Which poorly developed region of the brain would suggest that fish lack the capacity for higher reasoning skills that many mammals possess?

5. Match the structures on the left with the systems to which they belong on the right.

_____ gills
_____ liver
_____ pyloric ceca
_____ kidney
_____ gonad
_____ ureter
_____ telencephalon
_____ spleen
_____ bladder

a. reproductive
b. excretory
c. digestive
d. respiratory
e. circulatory
f. nervous

Amphibia

After completing the exercises in this chapter, you should be able to:

1. Identify the major skeletal features of the frog and compare them to homologous bones in other vertebrates.

2. Identify the major muscles of the frog and compare their morphology and functions to homologous muscles in other vertebrates.

3. Identify the major internal organs in the frog, discuss their functions in the body and relate them to homologous organs in other vertebrates.

4. Define all boldface terms.

One of the most significant events in vertebrate evolution was the gradual movement of small groups of early vertebrates from water onto land. Amphibians tell us a fabulous story about the characteristics that were present in these early vertebrates to allow such a successful shift. In many ways, modern amphibians represent an evolutionary transition between fish and terrestrial animals (reptiles, birds and mammals) and have a unique blend of characteristics well suited for life in and out of the water. As aquatic larvae, many possess the fusiform body shape and external gills of fish and have long, laterally-compressed tails to assist in swimming. Yet adults have a thinner, lightweight skull and jointed limbs for locomotion rather than fins. Lungs are present in most groups and the closed circulatory system is powered by a three-chambered heart which partially separates oxygenated blood from deoxygenated blood. The thin skin of amphibians lacks scales or other keratinized tissue and most amphibians are extremely susceptible to desiccation. Many also rely on the skin as a supplementary organ for respiration. As a result, most amphibians are found in damp, humid habitats that allow them to keep their skin sufficiently moist to prevent dehydration and facilitate gas exchange.

There are just over 4,000 currently living amphibian species which belong to the class Amphibia within the subphylum Vertebrata. In this chapter you will examine the anatomy of a frog as a representative amphibian. Keep in mind that no one amphibian could serve to accurately represent the many variations of body styles and characteristics within the class Amphibia, but frogs do possess most of the typical amphibian features. In addition, they display many unusual features that reflect specific adaptations to their own unique lifestyle.

EXERCISE 15–1

Photographic Atlas Reference Page 105

Amphibian Anatomy

Materials needed:
- mounted frog skeleton
- preserved frog
- dissecting tools
- dissecting pan

SKELETAL SYSTEM

1. Obtain a mounted, articulated skeleton of a frog.

2. Examine it carefully using Figure 15.1 to assist you in identifying the skeletal elements present.

Many of the bones seen in the frog represent homologies to bones present in most mammals, including humans (i.e., humerus, scapula, femur, atlas), providing evidence of the common ancestry of all vertebrates. However, many specializations in the frog skeleton reflect adaptations for jumping great distances and landing safely. Notice how the vertebral column is compressed into just 10 **vertebrae**. The tenth vertebra is the long **urostyle** which develops from the fusion of the last several vertebrae to form a strong, rigid base for bearing the forces necessary to propel the frog into a jump and to support its landing. Likewise, in the forelimbs and hindlimbs some bones are fused during development to increase strength in these areas of the body. The radius and ulna fuse to form a common **radioulna**, while the tibia and fibula fuse to form the **tibiofibula** (Fig. 15.1). The **tarsals** of each hindlimb are elongated and articulate with the tibiofibula and metatarsals to increase flexibility in the distal portion of the hindlimbs, enhancing both swimming and jumping.

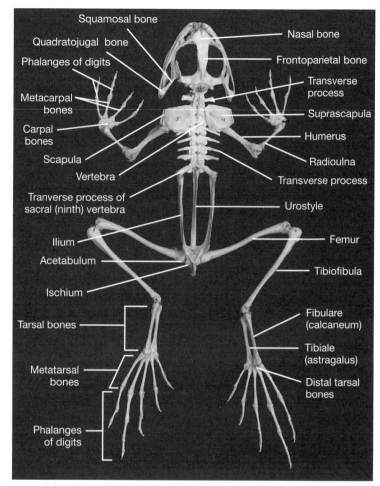

FIGURE 15.1 ▪ Dorsal view of the skeletal anatomy of a frog.

Check Your Progress

1. What evidence of segmentation do you see in the frog skeleton?

2. How many toes are present on the forelimbs? on the hindlimbs? How does this compare to mammals?

3. What bony remnant of the postanal tail that was present in the tadpole stage remains in the adult?

4. Are ribs present in the frog?

5. List 5 bones in the frog that are directly homologous to bones in humans.

MUSCULAR SYSTEM

With over 200 muscles, a comprehensive study of frog musculature would entail many hours of tedious dissection and is beyond the scope of this laboratory manual. Thus we will confine our study of the muscular system to the major muscles of the dorsal body surface. Since this is your first in-depth investigation of muscles, some background information should make your observations more productive.

All skeletal muscles have a fixed end, the **origin**, and a moveable end, the **insertion**. Most skeletal muscles taper from the thicker **muscle belly** in the middle to thinner, tough, white tendons on the ends which anchor the muscle to bones or other muscles. The direction a muscle exerts force (its **action**) also plays a role in its shape and where it inserts and originates. A muscle that **adducts** moves a limb toward the midline of the body. Conversely, a muscle that **abducts** moves a limb away from the midline of the body. Some muscles are **flexors** and bend one part of the body toward another part, while other muscles are **extensors** and straighten or extend a body part. Since muscles can contract in only one direction, and thus perform only one movement, they are generally arranged in **antagonistic** pairs or groups; that is, they work in opposition to one another. Because muscles are usually grouped closely together, it is often difficult to tell where one muscle ends and another begins. You should pay careful attention to the orientation of the muscle fibers. Often this will give you clues as to where two muscles cross or abut.

1. To observe the musculature of your frog, you must first remove the skin. Place the frog on its dorsal surface in your dissecting pan and make a small incision through the skin (but *not* through the body wall!) with scissors along the midventral line from the tip of the jaw caudally to the cloaca.

2. Start to separate the skin from the underlying muscles carefully, using a blunt probe where necessary.

3. Make two transverse incisions through the skin around the body, one in front of the forelimbs and the other just in front of the hindlimbs.

4. Use a blunt probe to carefully tease the skin away from the muscles around the entire body. In most

cases you should be able to "peel" the skin away with your fingers. Continue until you have completely exposed the muscles of your specimen.

5. Using a similar technique, remove the skin from the forelimbs and hindlimbs of your specimen.

6. Use Table 15.1 and Figure 15.2 to identify the major dorsal muscles and their actions in the frog.

TABLE 15.1 ▪ Superficial Musculature of the Frog

Muscle Name	Action
Temporalis	Flexes head and lifts mandible
Pterygoid	Raises head
Dorsalis scapulae	Adducts and stabilizes scapula
Deltoid	Flexes shoulder and adducts forelimb
Latissimus dorsi	Flexes shoulder and moves forelimb dorsally and cranially
Internal abdominal oblique	Compresses abdomen and flexes trunk
External abdominal oblique	Compresses abdomen and flexes trunk
Gluteus	Rotates thigh
Adductor magnus	Adducts thigh
Iliacus internus	Extends hindlimb
Peroneus	Flexes hindfoot
Tibialis anterior longus	Flexes hindfoot
Flexor digitorum brevis	Flexes phalanges of hindfoot
Abductor brevis dorsalis	Abducts phalanges of hindfoot
Gastrocnemius	Extends hindfoot
Gracilis minor	Extends thigh and flexes hindlimb
Semimembranosus	Extends thigh and flexes hindlimb
Biceps femoris	Flexes hindlimb
Triceps femoris	Flexes thigh and extends hindlimb
Longissimus dorsi	Extends vertebral column
Anconeus	Extends forearm
Extensor carpi ulnaris	Extends carpus (wrist)
Extensor digitorum communis	Extends phalanges
Extensor carpi radialis	Extends carpus (wrist)

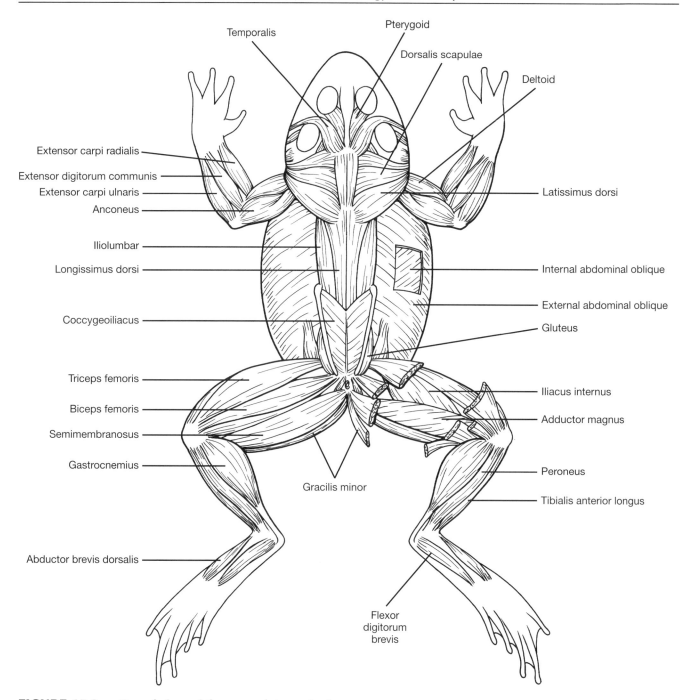

Temporalis

Pterygoid

Dorsalis scapulae

Deltoid

Extensor carpi radialis

Extensor digitorum communis

Extensor carpi ulnaris

Anconeus

Iliolumbar

Longissimus dorsi

Coccygeoiliacus

Triceps femoris

Biceps femoris

Semimembranosus

Gastrocnemius

Abductor brevis dorsalis

Gracilis minor

Flexor digitorum brevis

Latissimus dorsi

Internal abdominal oblique

External abdominal oblique

Gluteus

Iliacus internus

Adductor magnus

Peroneus

Tibialis anterior longus

FIGURE 15.2 ▪ Dorsal view of the musculature of a frog.

Photographic Atlas Reference Pages 109–112

INTERNAL ANATOMY

1. Place your frog on its dorsal surface in your dissecting pan.

2. Using scissors make a shallow, longitudinal incision along the midventral line through the body wall from the tip of the jaw caudally to the cloaca.

3. Make two transverse incisions along the sides of the body just behind the forelimbs and just in front of the hindlimbs.

4. Spread apart these flaps and pin them to the dissecting tray to view the internal organs of your specimen.

5. Use Table 15.2 and Figure 15.3 to identify these structures and their corresponding functions in the body.

Frogs capture prey in their mouths with the aid of their long, sticky tongues. Their mouths are wide to accommodate large prey items, but are not equipped for chewing; thus prey are swallowed whole. Food items pass down the **esophagus** and digestion begins in the **stomach**. The large, multi-lobed **liver** produces bile which is stored in the **gallbladder** and released into the **bile duct**. Bile, along with digestive enzymes from the **pancreas**, travels into the **duodenum**, the anterior portion of the small intestine. Nutrients are absorbed throughout the **small intestine** and water and ion absorption occur in the **large intestine** (Fig. 15.3). Indigestible food items exit the body through the **cloaca**. Small **lungs** are present behind the heart and liver and are used for respiration. No diaphragm is present in frogs, thus each breath must be "swallowed" and forced into the trachea using muscles in the throat region. The frog supplements gas exchange by use of highly-vascularized patches of skin which must remain moist. A small three-chambered **heart** is the driving force behind the closed circulatory system in amphibians. Separate pulmonary and systemic branches exist to route blood to the lungs and to the rest of the body, but oxygenated and deoxygenated blood are not kept completely separated as they are in mammals.

The **kidneys** are flat, elongated organs that lie on either side of the caudal vena cava dorsal to the intestinal tract. They have several arterial and venous attachments to the circulatory system along their length for blood filtration (Fig. 15.4). Metabolic wastes are filtered from the blood and passed through the **ureters** to the **bladder** for storage (Table 15.3). Frogs and toads are exceptionally good at conserving water and many species are able to directly reabsorb water from urine stored in their bladder during prolonged dry periods. Urine passes out of the frog through the **cloaca**, a common chamber for urine, feces and gametes. Male frogs possess a pair of testes—each is attached to the ventral surface of a kidney by **vasa efferentia** (Fig. 15.4). Sperm are transported through the vasa efferentia to the kidneys, then through the ureters to the cloaca for release from the body. Eggs are produced in the **ovaries** and are released into the **coelom** when mature (Fig. 15.5). They are captured by the funnel-shaped openings (the **ostia**) of each **oviduct** and transported to the corresponding **uterus** on that side. As the eggs pass along the length of the oviducts, they are encased with several layers of gelatinous material secreted by the oviductal walls. In frogs, the uteri are simply the enlarged, terminal portions of the oviducts that temporarily store unfertilized eggs prior to mating. As mating occurs, the female frog will release her eggs through the cloaca and the male frog will simultaneously release a cloud of sperm cells over them to complete the act of external fertilization.

TABLE 15.2 ▪ Internal Anatomy of the Frog

Structure	Function
Esophagus	Transports food to stomach
Liver	Produces bile and detoxifies many constituents of the absorbed digested compounds
Pancreas	Produces digestive enzymes and delivers them through common bile duct to duodenum
Stomach	Site of food storage and initiation of digestion
Spleen	Stores blood
Large intestine	Site of absorption of water as well as certain vitamins and ions
Ileum	Site of completion of digestion where most of the absorption of nutrients into the bloodstream occurs
Duodenum	Receives secretions from liver and pancreas through common bile duct for further breakdown of food from stomach
Bile duct	Transports secretions from the liver and pancreas to the duodenum
Gallbladder	Stores bile produced by the liver
Heart	Three-chambered organ (two atria, one ventricle) which pumps a portion of the venous blood to the lungs and then back to the heart before it is pumped to the various organ systems of the body through the systemic system

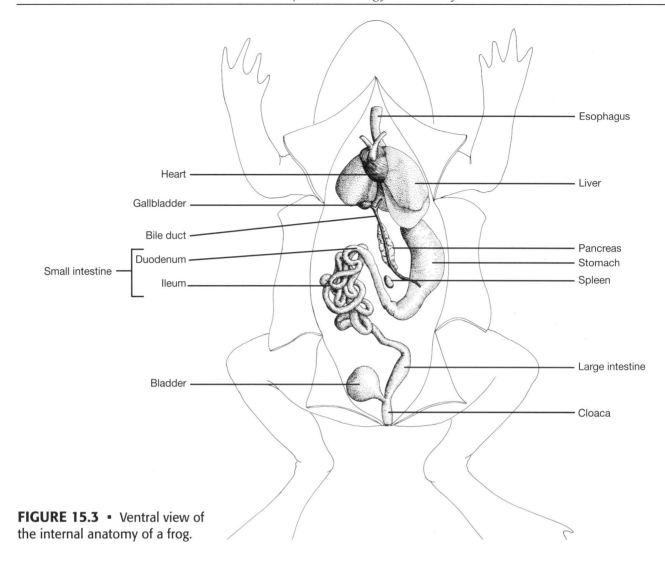

FIGURE 15.3 ▪ Ventral view of the internal anatomy of a frog.

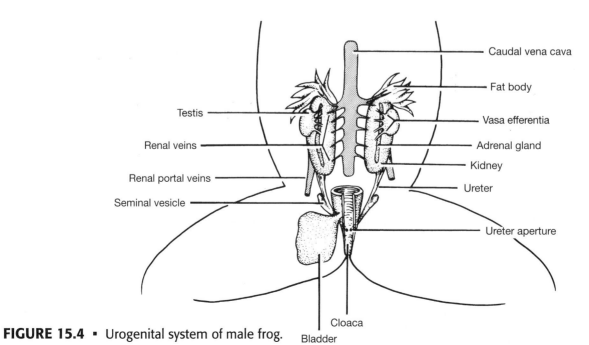

FIGURE 15.4 ▪ Urogenital system of male frog.

TABLE 15.3 · Abdominal Circulatory Vessels and Urogenital Organs of the Frog and Their Functions

Structure	Function
Caudal vena cava	Transports deoxygenated blood from the caudal regions to the sinus venosus
Fat bodies	Repositories for lipid reserves
Adrenal glands	Produce hormones which regulate blood pressure and metabolism
Kidneys	Paired organs which filter nitrogenous wastes from the blood
Renal veins	Transport filtered blood from the kidneys to the caudal vena cava
Renal portal veins	Transport blood from the caudal regions to the kidneys
Ureters	Transport urine from the kidneys to the bladder and in males, transport sperm to cloaca
Ureter apertures	Paired orifices through which urine from the ureters enters the cloaca before passing into the bladder for storage
Cloaca	Common chamber for collection of materials from the digestive, excretory and reproductive systems prior to their discharge from the body
Bladder	Site of storage for urine prior to discharge from cloaca
Vasa efferentia (male)	Small ducts that transport sperm from the testes to the kidneys
Seminal vesicles (male)	Contribute seminal fluid for sperm entering cloaca to assist in dispersion and insemination
Testes (male)	Paired organs which produce sperm for transport via the vasa efferentia through the kidneys to the ureters and discharge through the cloaca
Uteri (female)	Site of storage of unfertilized eggs prior to discharge through the cloaca
Ovaries (female)	Organs which produce eggs for transport through the oviducts to the uteri and discharge through the cloaca
Oviducts (female)	Paired tubes that transport eggs from the ovaries to the uteri and secrete jelly-like coating for protection of the eggs
Ostia (female) sing. = ostium	Openings in the cranial ends of the oviducts through which eggs released from the ovaries into the coelomic cavity enter the oviducts

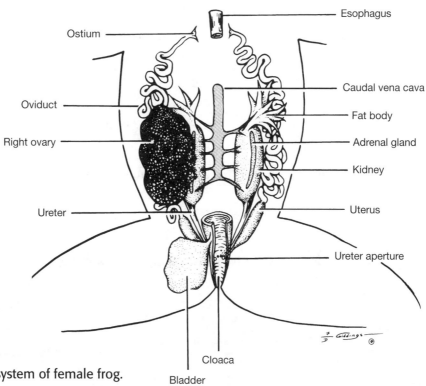

FIGURE 15.5 · Urogenital system of female frog.

Questions For Review

1. Which features of the skeletal system in frogs represent specific adaptations for jumping?

2. What is the functional difference between a cloaca and an anus?

3. How many chambers does a frog heart have? Is the circulatory system open or closed?

4. Which one of the following organs *is not* part of the female frog's reproductive pathway?
   a. ureter
   b. ovary
   c. oviduct
   d. uterus
   e. cloaca

5. List all of the organs in the male frog that are common to both the excretory *and* reproductive systems.

6. Which two accessory digestive organs release compounds into the duodenum through the bile duct?

7. What other organ, besides lungs, do frogs use to supplement gas exchange?

Reptilia

After completing the exercises in this chapter, you should be able to:

1. Identify the major external features of a turtle.

2. Identify the major internal organs of a turtle and a snake, discuss their functions in the body and relate them to homologous organs in other vertebrates.

3. Define all boldface terms.

In this chapter you will examine in detail two of the 6,550 species in the class Reptilia. The turtle and snake have been selected because, in one sense, they represent extremes at opposite ends of the evolutionary history of reptiles, yet they share many typical reptilian characteristics. Turtles have remained virtually unchanged for the past 200 million years, while snakes are a fairly recent evolutionary offshoot from the main branch of reptiles and have many highly specialized features. Reptiles were the first group of truly terrestrial animals with no dependence on water for reproduction. Reptiles have **direct development** with no larval stage.

All reptiles are characterized by the presence of dry skin comprised of overlapping, **keratinized scales.** In turtles, some of these scales have been modified into the bony scutes, or plates, of the shell. This tough outer skin of reptiles provides an excellent barrier against dehydration—a most formidable concern for all terrestrial animals. Unfortunately this type of skin allows for very limited growth and must be replaced on a continuous basis. Thus reptiles shed through a process known as **ecdysis.** Another feature that has contributed to the success of reptiles in terrestrial environments is the evolution of the **amniotic egg,** sometimes referred to as the cleiodoic egg. This type of egg has a tough (sometimes leathery), outer shell which is impermeable to water loss but permits gas exchange. For such an impenetrable barrier to be placed around an egg before it leaves the female's body, the egg has to be fertilized before this shell is constructed around it. Thus mechanisms of **internal fertilization** arose and with them, the dawn of **copulatory organs.**

Reptiles can also be classified as **ectotherms**—animals that derive the bulk of their internal body heat from external sources. Fish, amphibians and invertebrates all fall into this category. Only birds and mammals are true **endotherms** that generate the majority of their body heat from within.

EXERCISE 16–1

Photographic Atlas Reference Pages 115–118

Turtle Anatomy

Materials needed:
- preserved turtle
- dissecting tools
- dissecting pan

EXTERNAL ANATOMY

1. Obtain a preserved turtle and place it on its ventral surface in your dissecting pan.

2. Examine the paired forelimbs and hindlimbs. Your specimen is an aquatic species that has partially **webbed feet** to aid in swimming. Long **claws** are also present at the tips of the toes for digging and crawling.

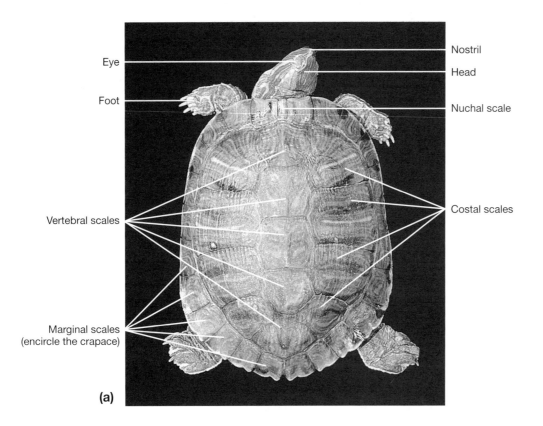

(a)

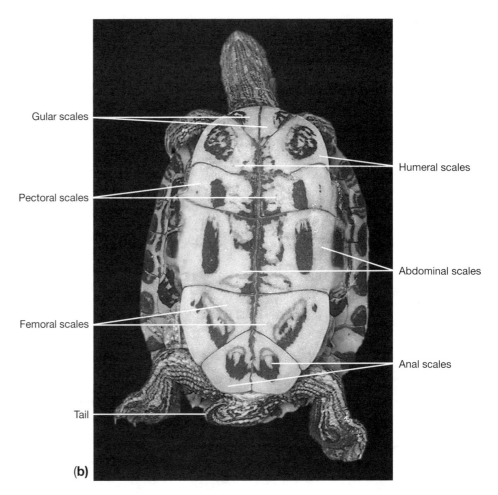

(b)

FIGURE 16.1 ▪ External anatomy of a turtle: (a) dorsal view and (b) ventral view.

3. Examine the head. **Eyes** and **ears** are present, although no external ear openings exist (Figure 16.1). Paired **nostrils** are also present at the rostral end of the snout. The tip of the mouth is covered by a sharp, epidermally-derived projection called the **beak**. Turtles lack teeth, but are quite capable of tearing apart their food into "bite-sized" pieces using their strong neck muscles and their sharp beaks.

4. Notice that space exists inside the shell at the base of the neck and each leg for the head and limbs to be withdrawn into the shell. This affords an extra degree of protection for the turtle when confronted with danger.

5. The most prominent feature of any turtle is without a doubt its shell. The shell is comprised of an upper section, the **carapace**, and a lower section, the **plastron**, that are fused along the lateral margins into one solid structure. In many species, the plastron has one or two transverse **hinges** which allow the ends of the plastron to open and close. This permits some turtles to tightly seal themselves inside their shells when threatened.

6. Examine the plastron. The hinges (when present) are located at the sutures along the cranial and caudal borders of the abdominal scales. Press against the ends of the plastron to determine the location of the hinges.

7. The **scutes** (scales) of the carapace and plastron are comprised of epidermal tissue covering hard, fused bony plates underneath. The names of the scutes generally correspond to the region of the body which they cover (i.e., vertebral scales, pectoral scales, abdominal scales). Biologists use the number, arrangement and color patterns of these scutes to identify different turtle species.

8. Although a turtle sheds its skin, its shell remains in place throughout the animal's life and thus bears the scars and signs of age that accumulate through the years. As was the case with fish scales, close examination of the scutes reveals concentric growth rings that give an idea of the relative age of the turtle. In general, each ring represents one year's growth.

9. If other turtle specimens are available, compare the number and arrangement of scutes on the plastron and carapace as well as the location and number of hinges on the plastron.

Check Your Progress

1. Count the growth rings on one of the costal scales on your specimen and estimate the turtle's age.

2. How many toes are present on the forelimbs? on the hindlimbs? How does this compare to mammals?

3. What features of the turtle head are homologous in mammals?

INTERNAL ANATOMY

1. Lay your turtle on its carapace with the plastron facing you.

2. Opening a turtle is a bit like opening a tin can in the middle without losing any of its contents. You must carefully separate the plastron from the carapace and other soft tissues along its entire margin.

3. Use a small saw to cut through the lateral margins of the shell and scissors or a scalpel to separate the plastron from the soft tissues of the neck and legs.

4. Use Table 16.1 and Figure 16.2 to guide you through this dissection.

The **esophagus** in most reptiles is equipped with several longitudinal folds to permit great distensibility for swallowing large food items. The **stomach**, found beneath the right lobes of the liver, resembles that of a mammal quite closely, as do the duodenum and convoluted small intestine. The **gallbladder** may be found underneath the left lobe of the liver. Its function is the same as in other vertebrates. A **cecum** is present in turtles (and some snakes) for fermentation. The **cloaca** of reptiles is divided internally into three chambers (as in birds). The first chamber is the **coprodeum** which receives fecal matter directly from the intestine. The middle chamber is the **urodeum** which receives reproductive and urinary products, and the last chamber is the **proctodeum** which acts as a general collecting area for digestive and excretory wastes.

The respiratory system of turtles follows the typical vertebrate plan with a **trachea** branching into two primary **bronchi**, one leading to each **lung**. Two functional lungs are present, and in turtles and crocodiles the cartilaginous rings completely surround the trachea preventing collapse of this structure during respiration or feeding. Reptiles (with the exception of crocodiles) lack a diaphragm and must use the intercostal muscles of the ribs or the movement of other visceral organs to breathe. Reptiles have a **three-chambered heart** that is somewhat more sophisticated in its internal design than the amphibian heart and, despite its single ventricle, basically functions as a four-chambered heart keeping oxygenated and deoxygenated blood sufficiently separated. A usual feature of the reptilian circulatory system is the **renal portal system**. Blood from the tail and rear legs passes through the kidneys before returning to the systemic circulatory system. A small **spleen** is present and usually found adhering tightly to the **pancreas**.

Reptiles have a metanephric **kidney** that is more advanced than fish and amphibian kidneys and is more adept at conserving water. The kidneys are

TABLE 16.1 ▪ Internal Anatomy of the Turtle

Structure	Function
Trachea	Conducts air to and from lungs during respiration
Thyroid gland	Controls metabolism, growth rates, blood calcium levels and shedding cycle
Liver	Produces bile, converts glucose to glycogen for storage, detoxifies many constituents of the absorbed digested compounds
Gallbladder	Stores bile produced by the liver
Heart	Three-chambered organ (two atria, one ventricle) that receives oxygenated blood from the lungs and pumps it via the arteries throughout the body
Lungs	Paired, highly-vascularized organs for respiration
Stomach	Site of food storage and initiation of digestion
Large intestine	Responsible for reabsorption of water and electrolytes; transports feces to coprodeum via peristalsis
Cecum	Large, thin-walled pouch in herbivorous species demarcating the beginning of the large intestine; contains anaerobic bacteria responsible for fermentation of cellulose and other plant materials
Pancreas	Produces digestive enzymes and delivers them through pancreatic duct to duodenum; also produces a suite of hormones for the endocrine system
Spleen	Stores blood, recycles worn-out red blood cells, produces lymphocytes
Kidneys	Filter blood (creating urine) and responsible for osmoregulation
Testes (male)	Produce sperm
Ovaries (female)	Produce eggs

typically offset in the turtle with the right kidney in front of the left. Urine enters the urodeum (of the large intestine) from the **ureters** and then flows backward into the bladder for storage. Turtles have a large **urinary bladder** and can re-circulate urine through their bodies to reuse the water in it. The location of the gonads varies slightly in males and females. In males, the **testes** are attached to kidneys. In females, the **ovaries** are attached to the caudal side of the lungs near the rear of the shell. As eggs mature in the ovaries, they pass into the **shell gland** (uterus) where the eggs are fertilized and subsequently encased in their protective, leathery shells. Male turtles possess a single copulatory organ, the **penis**, which is housed in the front portion of the cloaca within the base of the tail and everts into the proctodeum.

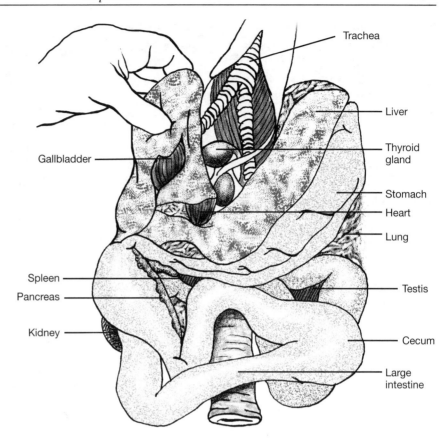

FIGURE 16.2 ▪ Ventral view of the internal anatomy of male turtle.

Check Your Progress

1. Are turtles dioecious?

2. How many functional lungs do turtles possess?

3. How is the reptilian uterus modified for increasing the survival of the eggs?

4. Would you expect the cecum to be relatively large or small in herbivorous turtles?

Snake Anatomy

Materials needed:
- preserved snake
- dissecting tools
- dissecting pan

EXTERNAL ANATOMY

The fact that snakes have lost their legs over the course of evolution has actually improved their ability to move. Some snakes, such as racers and coach-whips, can easily outdistance humans over difficult terrain. Eyelids in snakes are also absent; instead a transparent scale called the **spectacle** protects the eye. Ears are absent as well. Most snakes respond to low frequency vibrations that they detect through the ground. What snakes lack in their ability to hear they make up for many times over in their exceptional ability to smell. Snakes have a special sensory structure called the **vomeronasal organ** (Jacobson's organ) that has paired openings in the roof of the mouth. As the tongue flicks, it picks up scent molecules on each fork which are deposited into separate openings of the vomeronasal organ. In fact, the nerve fibers from each opening remain separate all the way to the brain, giving snakes the ability to smell in stereo! By simply probing the air or substrate with their tongues, they are able to determine the direction and proximity of objects by detecting minute differences in the relative concentrations of odor molecules on each fork of the tongue. Snakes possess **teeth**, but do not chew their prey. Non-venomous snakes have four rows of upper teeth and two rows of lower teeth. The teeth are recurved and angle toward the back of the mouth to facilitate swallowing. In venomous snakes a pair of maxillary teeth are usually modified into hollow, surgically-sharp fangs to deliver their toxic venom deep into the tissue of their prey.

INTERNAL ANATOMY

1. Use dissecting scissors to make parallel, longitudinal incisions along each side of the body, starting at the cloacal opening and progressing cranially the entire length of the body to the base of the head.

2. Make transverse incisions across the body at the ends of these incisions to connect them and carefully remove the skin from the ventral surface of your specimen.

3. Use Table 16.2 and Figure 16.3 to help you identify the internal organs in the snake.

For practical purposes, it is easiest to mentally divide the snake body into thirds. The **trachea, esophagus** and **heart** are the primary organs in the first third of the body. In the second third you will find the **liver, stomach, pancreas, right lung** and **gallbladder**. The stomach is tube-like and rather short in snakes. In the most caudal third you will find the **small intestine, cecum, large intestine, kidneys, gonads** and **cloaca**.

The respiratory system of snakes is rather unique. The long **trachea** contains the cartilaginous rings

TABLE 16.2 ▪ Internal Anatomy of the Snake

Structure	Function
Trachea	Conducts air to and from lung during respiration
Thyroid gland	Controls metabolism, growth rates, blood calcium levels and shedding cycle
Heart	Three-chambered organ (two atria, one ventricle) that receives oxygenated blood from the lungs and pumps it via the arteries throughout the body
Lung	Single, highly-vascularized organ for respiration; right lung is functional while left lung is reduced to a vestigial remnant
Liver	Produces bile, converts glucose to glycogen for storage, detoxifies many constituents of the absorbed digested compounds
Air sac	In some water snakes, the left lung develops into a gas-filled chamber that enhances buoyancy and exerts force against the body wall to counteract water pressure
Stomach	Site of food storage and initiation of digestion
Spleen	Stores blood, recycles worn-out red blood cells, produces lymphocytes
Gallbladder	Stores bile produced by the liver
Pancreas	Produces digestive enzymes and delivers them through pancreatic duct to duodenum; also produces a suite of hormones for the endocrine system
Ovaries	Produce eggs
Oviducts	Capture eggs and transport them to shell gland (uterus)
Kidneys	Filter blood (creating urine) and responsible for osmoregulation
Large intestine	Responsible for reabsorption of water and electrolytes; transports feces to coprodeum via peristalsis
Cloaca	Common chamber for the release of urine, feces and reproductive products

present in most terrestrial vertebrates, but the rings are incomplete and do not entirely surround the trachea. This permits the trachea to collapse somewhat to allow large food items to pass through the expandable **esophagus** more easily. Snakes have only one functional lung, the **right lung**. In most snakes the left bronchus terminates in a vestigial, nonfunctional lung. In some water snakes this nonfunctional lung has been modified into an **air sac** (Table 16.2). Like most reptiles, snakes lack a diaphragm. Circulation follows the typical pattern seen in reptiles with a three-chambered heart powering a closed circulatory system with arteries and veins. The **renal portal system**, discussed earlier, is also present in snakes.

Snakes lack a urinary bladder; thus urine storage is minimal and is confined to the small volume of fluid that the **ureters** can hold. As in turtles and lizards, urine flows from the ureters into the **urodeum** and out of the body through the cloaca. Paired gonads are located just cranial to the kidneys. Male snakes and lizards possess paired reproductive structures called **hemipenes**. Each hemipenis is attached directly to one of the **testes**. Only a single hemipenis is used when mating, but snakes (and lizards) often alternate between the left and right hemipenis when successive matings occur within a short time interval. This behavior maximizes the number of viable sperm that can be released with each mating attempt.

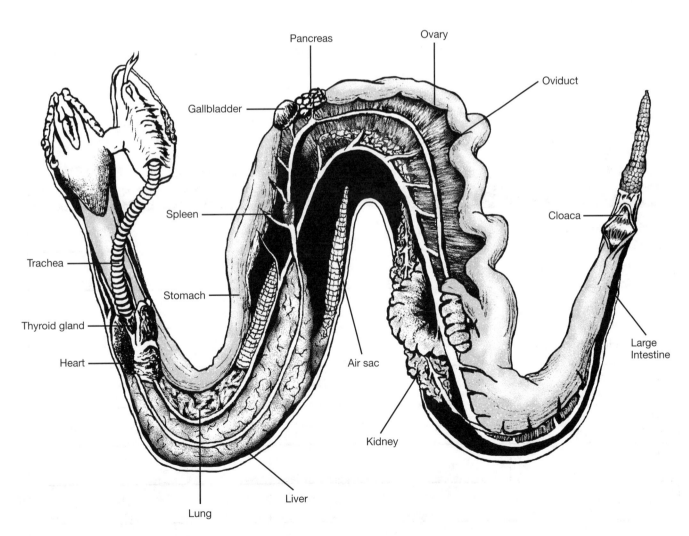

FIGURE 16.3 ▪ Ventral view of the internal anatomy of female snake.

 **Questions For Review**

1. Name four characteristic features of reptiles that have directly contributed to their success in terrestrial habitats.

   a.

   b.

   c.

   d.

2. Define ectothermic.

3. How many chambers does a turtle's heart have? Is the circulatory system open or closed?

4. In addition to regulating metabolism, growth and blood calcium levels, what additional function does the thyroid gland perform in reptiles?

5. Fill in Table 16.3 with the appropriate characteristics for each group of reptiles to compare the adaptations to different body shapes and lifestyles between turtles and snakes.

TABLE 16.3 ▪ Comparison of Anatomical Characteristics Between the Turtle and the Snake

	Turtle	Snake
Number of limbs		
Eyelids (present or absent)		
Teeth (present or absent)		
Number of functional lungs		
Tracheal rings (complete or partial)		
Diaphragm (present or absent)		
Urinary bladder (present or absent)		
Location of urine storage		
Number of copulatory organs in male		

Aves

After completing the exercises in this chapter, you should be able to:

1. Identify the major skeletal features of the pigeon and compare them to homologous bones in other vertebrates.

2. Identify the major muscles of the pigeon and compare their morphology and functions to homologous muscles in other vertebrates.

3. Identify the major internal organs in the pigeon, discuss their functions in the body and relate them to homologous organs in other vertebrates.

4. Define all boldface terms.

Birds are the most numerous of all terrestrial vertebrate groups. The class Aves contains nearly 8,800 different species. Although birds and reptiles appear drastically different, these two groups share a very recent evolutionary past. The first birds diverged from the main reptilian line around 150 million years ago. Yet in that brief time span, they have diversified and adapted to fill many niches that reptiles and mammals were not equipped to exploit. Birds and mammals are the only groups of **endothermic** animals, meaning that they generate the majority of their body heat from within. Birds and mammals are often referred to as **homeothermic** organisms, meaning that their internal core body temperatures do not fluctuate much from their set points. The hallmark of birds is their magnificent **feathers** which provide protection, insulation and a lightweight surface area for flight. Birds generally have large, well-developed **eyes** and correspondingly large visual centers in the brain. **Ears** are present but are hidden beneath the plumage of the head. Their narrow jaws form a horned **beak** that lacks teeth and superficially resembles the beaks of many turtles. Their bones are hollow to reduce weight and show marked fusion, especially in areas that must bear the forces generated in flight and while landing. Birds possess a **four-chambered heart** with separate pulmonary and systemic circuits, much like the heart of mammals.

| EXERCISE 17-1 | Photographic Atlas Reference Pages 129–131 |

Avian Anatomy

Materials needed:
- mounted bird skeleton
- preserved pigeon
- dissecting tools
- dissecting pan

SKELETAL SYSTEM

1. Obtain a mounted, articulated skeleton of a pigeon.

2. Examine it carefully using Figure 17.1 to assist in identifying the skeletal elements present.

Many adaptations for flight are apparent in the skeleton of birds, the primary adaptation being **hollow bones**. Birds have extremely thin, lightweight bones, many of which bear pneumatic extensions from the lungs. Thus a bird's skeleton constitutes a much smaller percentage of the animal's total body weight than does a mammal's skeleton. Notice the long, articulated vertebrae of the neck. They are organized to provide extreme mobility and flexibility of the head so that it remains fixed in position during flight, while the torso oscillates freely with each wing beat. A flexible neck also aids in feeding and preening. In sharp contrast to the long neck, the vertebrae of the trunk are compressed into a strong, fused group of bones that serves as a solid attachment for the large muscles of the pectoral and pelvic regions. The **synsacrum** represents 13 posterior vertebrae fused to bear the animal's weight on its two legs, while the **pygostyle** represents several fused

caudal vertebrae that support the tail feathers. Along the midventral line of the torso, the **sternum** has a prominent **keel** for the attachment of the large pectoralis muscles which drive the wings during flight. Most of the bones of the legs and wings are homologous with limb bones in other vertebrates, though significant modifications of the basic tetrapod blueprint have been made during the course of avian evolution. With respect to the legs, the fifth toe has been lost and the first toe is pointed caudally, acting as a "prop" on the ground and allowing birds to perch on tree limbs. The many bones of the ankle are fused into a single **tarsometatarsus** that acts as a shock absorber during landings. The location and articulation of each femur with the pelvic girdle helps shift the weight of the animal forward, keeping its center of gravity over the feet. The wings bear three short fingers and a fused **carpometacarpus**.

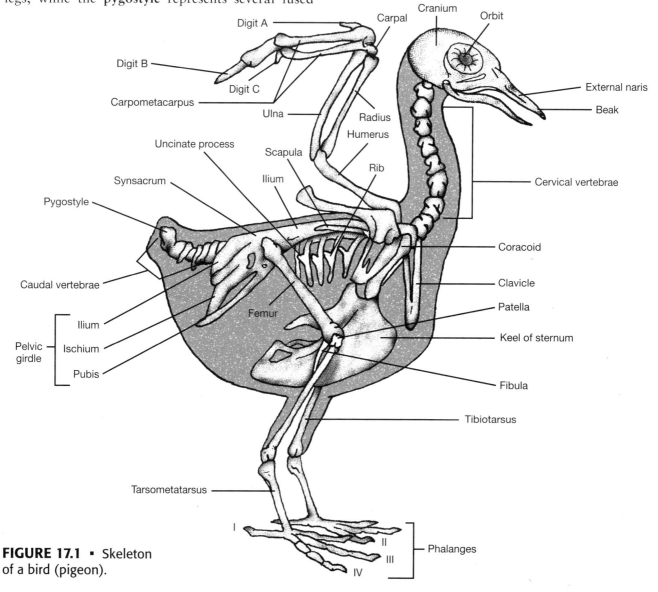

FIGURE 17.1 ▪ Skeleton of a bird (pigeon).

Check Your Progress

1. What major feature of bird bones has made flight possible?

2. List 5 bones in the pigeon that are directly homologous to bones in humans.

3. List 5 bones in the pigeon that do not have direct homologues in humans.

MUSCULAR SYSTEM

Due to the modifications of the bird's body for flight, many muscles in birds that are homologous in origin to other vertebrate muscles (and thus have similar names) have changed in their functions (e.g., the extensor metacarpi radialis *flexes* the carpometacarpus, instead of extending it as its name implies). In addition, many of the muscles are small and difficult to dissect, given the time frame of a single lab period. Therefore we will confine our investigation of avian musculature to the major superficial muscles of the wing, trunk and leg. Muscles account for a significant percentage of a bird's total mass—anywhere from 30% to 60%. The relative distribution of these muscles is correlated with the lifestyle of the species. Strong fliers such as pigeons, doves and hummingbirds have relatively heavy flight muscles and light leg muscles, while birds that rely more on their legs, such as ostriches, have considerably heavier leg muscles.

1. First remove the feathers and skin from the trunk, wing and leg regions of your specimen. The feathers and skin may be removed together, but be careful that the smaller muscles in the wing area are not destroyed in the process.

2. To save time, you may opt to dissect only the musculature on the right side of the bird's body.

3. Use Figure 17.2 as a guide for the dissection and identification of these muscles.

4. The actions of the major superficial muscles of the pigeon are listed in Table 17.1.

Clearly the largest muscle in the pigeon's body is the **pectoralis muscle**—the muscle that provides the powerful downward thrust of the wing. Together, the left and right pectoralis muscles average 15% of the total body mass in these birds. The mobility of the wings is due to the humerus being able to move in any direction. The elbow and wrist joints, conversely, are fairly inflexible, giving the wing considerable vertical stiffness. This also allows the wing to open and close as a unit during flight or displays. Leg musculature is concentrated toward the proximal portion of the leg to keep weight closer to the center of the body. A secondary benefit of this placement is that leg muscles can be adequately insulated by feathers, rather than lying exposed on the naked ends of the legs. Almost all of the actions of the foot and toes are controlled by muscles located in the proximal portion of the leg through a complex series of tendons which stretch along the tibiotarsus and tarsometatarsus. In perching birds, bending of the ankle joint automatically causes flexion of the toes, allowing a secure grip without the exertion of continuous muscular force.

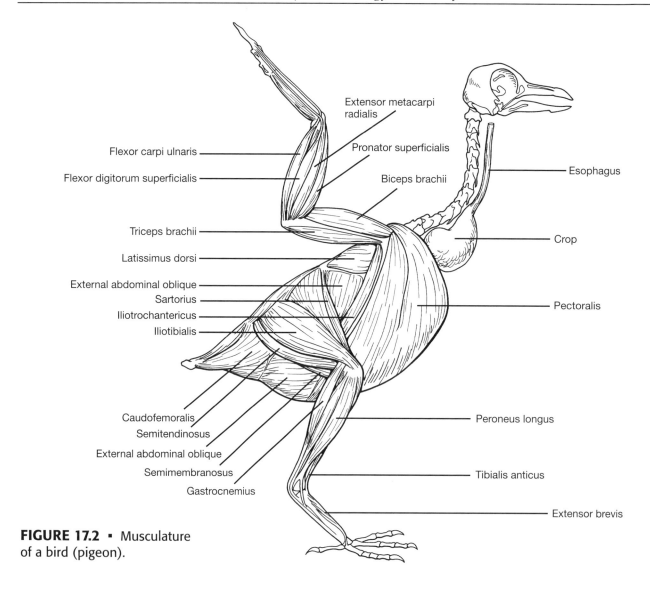

FIGURE 17.2 ▪ Musculature of a bird (pigeon).

TABLE 17.1 ▪ Selected Superficial Musculature of the Pigeon

Muscle Name	Action	Muscle Name	Action
Extensor metacarpi radialis	Flexes the carpometacarpus	Pectoralis	Depresses wing
Flexor carpi ulnaris	Flexes the carpometacarpus	Sartorius	Extends leg, adducts thigh
Flexor digitorum superficialis	Flexes the second digit	Iliotrochantericus	Abducts thigh
Pronator superficialis	Pronates the distal portion of the wing	Iliotibialis	Retracts or abducts the femur and extends the tibiotarsus
Biceps brachii	Flexes the forearm portion of the wing	Caudofemoralis	Abducts and extends thigh
Triceps brachii	Extends the forearm portion of the wing	Semitendinosus	Extends thigh and flexes leg
		Semimembranosus	Draws thigh caudally
Latissimus dorsi	Pulls humerus medially	Gastrocnemius	Extends tarsometatarsus
		Peroneus longus	Flexes digits of foot
External abdominal oblique	Supports and compresses abdominal body wall during exhalation	Tibialis anticus	Extends and abducts the knee while flexing the tarsometatarsus
		Extensor brevis	Extends digits of foot

Check Your Progress

1. Based on the relative distribution of musculature in the pigeon, would you guess that pigeons rely more on their legs or their wings?

2. Why are there very few muscles on the lower (distal) portions of the legs?

3. Compare the musculature of the frog (Fig. 15.2) with the musculature of the pigeon (Fig. 17.2) and list as many homologous muscles as you can find.

INTERNAL ANATOMY

NOTE: Dissection of the air sacs should be performed before dissection of the visceral organ systems using specially prepared specimens with latex injection of the air sacs and spaces. Check with your instructor to see if a demonstration specimen is available with the respiratory system already dissected.

1. Position your specimen on its back with its ventral surface facing you and use a scalpel to cut transversely through the middle of the pectoralis muscle and underlying muscles. Remove the musculature from the sternum.

2. Use scissors to cut through the keel of the sternum and separate the underlying membranes from the bone.

3. Make an incision along the midline of the abdominal wall to open the abdominal cavity.

4. Use scissors or bone cutters to cut through each rib near its articulation with the vertebrae. You may need to completely remove one of the wings to gain access to the lungs and air sacs.

5. Use Figure 17.3 to identify the internal anatomy of the bird. The functions of all organs depicted in Figure 17.3 are described in Table 17.2.

Most birds possess **external nares** for breathing and air travels down the **trachea** to the small, paired **lungs**. In mammals the trachea bifurcates over 20 times, yielding more than a million tubules which terminate in the thin-walled alveoli where gas exchange occurs. In birds, however, the trachea divides only a few times, forming two main bronchi which lead to the lungs and then branch into only a few dozen **secondary** and **tertiary bronchi** within the lungs. The **air sacs** constitute about 80% of the total volume of the respiratory system in birds and may completely surround the heart, liver, kidneys, testes, ovaries and intestines. Generally each air sac has one connection to a primary or secondary bronchus, but it may have several indirect connections to tertiary bronchi. Gas exchange occurs along tiny air capillaries which emanate from each bronchus. The **syrinx**, or voice box, is located at the caudal end of the trachea and is responsible for the large sound repertoire of most birds.

In mammals the entire lung expands and contracts with assistance from the diaphragm during each inhalation and exhalation. Birds lack a diaphragm and, thus, rely on movements of the ribs and sternum to bring in and expel air. Although air flows into and out of the trachea and nostrils tidally, air flows through the lungs and air sacs in a nearly unidirectional path during both inhalation and exhalation. The benefit of this phenomenon is twofold: (1) no dead air space exists in the avian respiratory system and (2) "fresh" air is passing over some portion of the respiratory tract at all times—with every inhalation and exhalation. The mechanism behind this feat is still not clearly understood, and several hypotheses have been advanced as to how birds attain this degree of efficiency. Collectively, the unique anatomy and bizarre physiology of the avian respiratory system make it one of the most complicated and poorly understood respiratory systems in the animal kingdom.

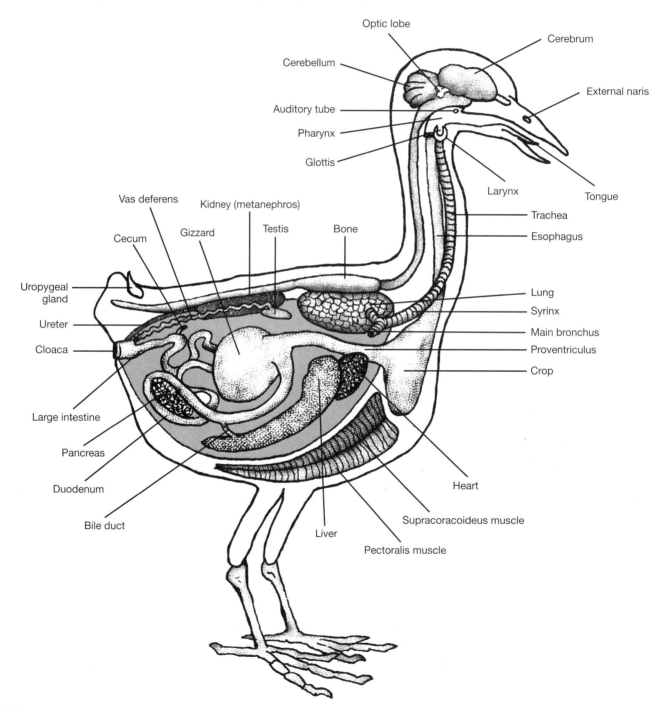

FIGURE 17.3 ▪ Internal anatomy of a bird (pigeon).

Since birds lack teeth, mechanical digestion of food begins in the **gizzard**. The **tongue** lacks taste buds and is used primarily to manipulate food into the mouth and down the **esophagus** into the crop. The **crop** stores food and regulates the passage of food into the gizzard. Food passes from the crop to the gizzard via a short tube known as the **proventriculus**. This digestive passageway mixes peptic enzymes with the food to begin the process of chemical digestion. The **gizzard** then thoroughly pulverizes and churns the food before sending it into the intestinal tract. A **liver** and **pancreas** are present and

TABLE 17.2 ▪ Internal Anatomy of the Pigeon

Structure	Function
External nares	Paired openings in beak for breathing
Tongue	Used to manipulate food into the mouth
Larynx	Opens and closes glottis during respiratory cycle and prevents foreign material from entering the lower respiratory tract
Trachea	Conducts air to and from lungs during respiration
Esophagus	Transports food to crop
Lungs	Paired organs of respiration connected to air sacs permitting unidirectional flow of air
Syrinx	Structure at the caudal end of the trachea responsible for calls and vocalizations
Main bronchi	Conduct inspired air both to the lungs and to the posterior air sacs
Crop	Stores ingested food to await passage to gizzard; found in seed- and grain-eating species
Gizzard	Thick-walled, muscular pouch which pulverizes and churns food prior to its passage into the intestine; larger in seed-eating species
Heart	Four-chambered organ (two atria, two ventricles) that receives oxygenated blood from the lungs and pumps it via the arteries throughout the body
Bile ducts	Transport bile from liver directly into the small intestine
Pancreas	Produces digestive enzymes that are released into the small intestine
Cecum	Vestigial structure in birds that is homologous to the appendix in humans
Large intestine	Continued digestion of food particles; transports feces to coprodeum of cloaca via peristalsis
Cloaca	Common chamber for collection of materials from the digestive, excretory and reproductive systems prior to their discharge from the body; partitioned into a coprodeum, a urodeum and a proctodeum as in reptiles
Ureters	Transport urine from the kidneys to the bladder
Uropygeal gland	Secretes oily compound that coats feathers, preventing them from getting wet
Vasa deferentia	Transport sperm from the testes to urodeum of cloaca
Kidneys (metanephros)	Paired organs which filter nitrogenous wastes from the blood and produce dry uric acid with little water that is transported to the urodeum of the cloaca where it mixes with feces as it is released from the body
Testes (male)	Produce sperm
Ovary (female)	Single organ (left only) which produces eggs
Proventriculus	Mixes food with peptic enzymes for digestion
Glottis	Prevents food from passing into trachea as it is swallowed
Pharynx	Region of oral cavity common to both respiratory and digestive pathways
Auditory tube	Channels sound waves to auditory receptors; similar to the ear canal of humans
Cerebellum	Primarily a reflex center for the integration of skeletal muscle movements; responsible for coordination and balance
Optic lobes	Process visual information from the eyes
Cerebrum	Largest portion of the brain; interprets sensory impulses and coordinates voluntary movements

release their respective digestive compounds through multiple ducts into the **small intestine** through which nutrients are absorbed into the bloodstream. A vestigial **cecum** is present and the **large intestine** finishes the digestive process and transports feces to the coprodeum of the **cloaca**. As in reptiles, the cloaca is partitioned into three semi-distinct chambers: the **coprodeum**, the **urodeum** and the **proctodeum**.

In birds, the female reproductive organs are limited to the left side. The right ovary and oviduct degenerate during development and occasionally an oviductal remnant may be found attached to the cloaca. In males, both **testes** are present and increase substantially in size as the breeding season approaches (up to 1000 fold!). They are located in the abdomen along either side of the dorsal aorta and are usually surrounded by air sacs. Movement of air through the air sacs is presumably responsible for cooling the testes to maximize sperm production (since optimal sperm production occurs at temperatures lower than the average body temperature of most birds). Sperm travel along the **vasa deferentia** to the **urodeum**, from which they are released from the body. Pigeons lack a penis, as do most birds, and the male deposits sperm in the female by pressing his **cloaca** tightly against hers as he ejaculates. (Domestic waterfowl such as chickens, turkeys, ducks and geese have a penis which channels sperm directly into the urodeum of the female.) As in reptiles, fertilization is internal and eggs are shelled by secretions of the oviductal walls. The **vagina** opens into the ventral surface of the **urodeum**, through which eggs are laid. Unlike in mammals, the sex of the offspring in birds is determined by chromosomes in the egg rather than in the sperm.

Excretion in birds is accomplished by the kidneys and the salt glands. Avian kidneys have some features resembling reptilian kidneys and other features that are more similar to the kidneys of mammals. The **kidneys** are paired and consist of three rather distinct lobes. There is no definitive medullary region as in mammalian kidneys (instead there are numerous medullary cones within each lobe) and the **ureters** that drain the kidneys open directly into the **urodeum** of the cloaca. No urinary bladder is present in birds. Nitrogenous wastes are excreted as **uric acid** (as in many reptiles), and these wastes mix with fecal matter from the coprodeum before being voided from the body. In addition to the kidneys, all birds possess paired **salt glands** embedded in the orbits of the eyes. In marine birds these glands are especially well-developed and may contribute up to 90% or more of the salts excreted by the body. The viscous salty fluid, composed principally of sodium chloride, is released through small secretory pores located near the eyes or along the beak.

Check Your Progress

1. Functionally, how has the larynx in humans been modified differently than in birds?

2. Number the following organs in the order that represents the correct path of food through the digestive tract of a bird:

   _____ cloacal opening

   _____ small intestine

   _____ crop

   _____ proventriculus

   _____ mouth

   _____ large intestine

   _____ esophagus

   _____ gizzard

   _____ coprodeum

3. Which side of the female reproductive tract is functional in birds?

4. Do birds possess a urinary bladder?

Questions For Review

1. Name at least two structures in the avian respiratory system that are not present in mammals.

2. What unusual excretory organ do birds possess on their heads?

3. What organ in birds secretes an oily compound to waterproof the feathers?

4. How many chambers does a bird's heart possess? Is this more similar to the heart of reptiles or to that of mammals?

5. Do birds possess a diaphragm? How is ventilation of the lungs accomplished?

6. What anatomical features of birds permit them to maintain constant and relatively high internal body temperatures?

7. List the three subregions of the avian cloaca.

8. Identify several anatomical similarities between birds and reptiles that reflect the common evolutionary lineage of these two groups.

Mammalia

After completing the exercises in this chapter, you should be able to:

1. Identify the major anatomical structures of the rat, discuss their functions in the body and compare them to homologous structures in other vertebrates.

2. Define all boldface terms.

The earliest mammals probably arose sometime around 190 million years ago and were contemporaries of the dinosaurs for 125 million years. During that time mammals were small, secretive, nocturnal creatures that were relegated to the few ecological niches that the large, ectothermic dinosaurs could not dominate. Around 65 million years ago, as the reign of the dinosaurs came to an abrupt end, mammalian evolution exploded and diversification occurred, driven by adaptations to the newly opened niches left vacant by the massive Cretaceous extinctions.

Today, the nearly 4,500 species within the class Mammalia possess a suite of common characteristics that sets them apart from all other chordates. Mammals are **endothermic** organisms that have epidermally derived **hair** covering their bodies for insulation. All but the spiny anteater and platypus are **viviparous**, meaning they give birth to live young. The male intromittent organ (**penis**) permits **internal fertilization** and the embryo is retained within the uterus of the female for the duration of fetal development.

Although mammals produce relatively few young, they invest considerable time and effort in caring for them. In females modified sweat glands, known as **mammary glands**, produce and secrete milk to nourish young. In most mammalian groups, fetal membranes are present (i.e., allantois, chorion, amnion) and the fetus receives nutrients and oxygen from its mother through its **placental attachment**. Mammals are characterized by **heterodont dentition** —teeth that differ structurally to accommodate foods in different ways.

Perhaps the most universal, yet most subtle, characteristic of mammals is that the lower jaw is comprised of a single bone, the **dentary bone**. A muscular **diaphragm** is present, separating the thoracic and abdominal cavities and assisting with ventilation of the lungs, and a **four-chambered heart** is present with separate pulmonary and systemic circuits keeping oxygenated and deoxygenated blood apart. A characteristic of mammals that played a major role in their evolutionary rise to the pinnacle of the animal world is the relatively large cerebrum present in most groups. This feature, coupled with superior sensory structures, gives mammals the ability to detect and respond to stimuli, modify their behaviors, and communicate with other members in their group.

Most mammals are terrestrial, although whales and porpoises are highly adapted mammals designed for aquatic living. Mammals also display tremendous diversity in size, ranging from small bats weighing only 1.5 grams to giant whales topping the scales at over 200,000 pounds!

Mammalian Anatomy

Materials needed:
- preserved rat
- dissecting tools
- dissecting pan

EXTERNAL ANATOMY

1. Obtain a preserved specimen of a rat and use Figure 18.1 to identify the external features of your specimen. These structures are defined in Table 18.1.

2. Most mammals follow the basic body plan of tetrapods, having two forelimbs and two hindlimbs. Rats use all four limbs for locomotion and thus have similar anatomical features on all limbs.

3. Determine the sex of your specimen. Males are recognized by the **scrotum** descending from the ventral portion of the abdomen near the tail (Fig. 18.1a). A **preputial orifice** housing the **penis** is located on the ventral surface of the abdomen between the thighs. Females will have a small opening, the **vulva**, just in front of the **anus**. **Mammary papillae** are often present in both sexes of mammals, but are more prominent and only functional in females (Fig. 18.1b).

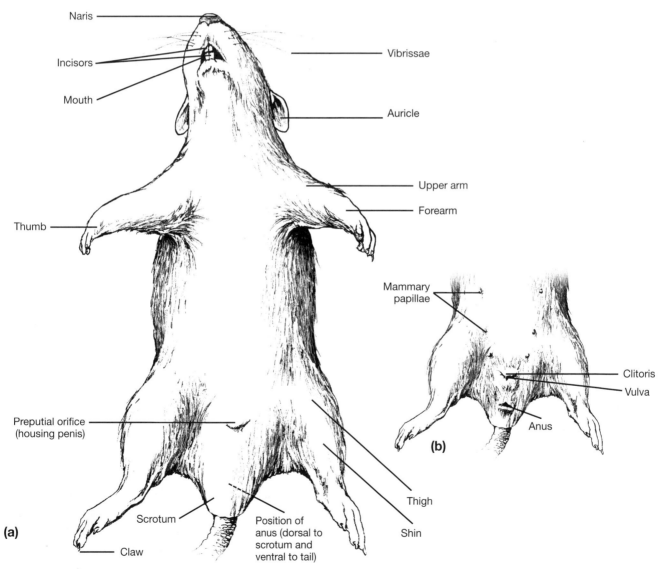

FIGURE 18.1 ▪ External anatomy of (a) male and (b) female rat.

TABLE 18.1 ▪ External Anatomy of the Rat

Structure	Function
Nares (sing. = naris)	Openings in nose for breathing
Vibrissae	"Whiskers" that respond to tactile stimuli
Incisors	Anterior teeth used to cut and gnaw
Mouth	External opening to oral cavity
Auricles	Dish-shaped external ears for collection of sound waves
Preputial orifice (male)	External opening to prepuce (sheath) through which penis extends
Scrotum (male)	Pouch extending from the caudal region that houses the testes
Mammary papillae (female)	Small protuberances on the ventral body surface through which the mammary glands secrete milk
Clitoris (female)	Homologous structure to the penis which plays a role in sexual sensation and stimulation
Vulva (female)	Most caudal region of the female urogenital tract
Anus	Regulates egestion of undigested food (feces) from the body

Check Your Progress

1. What external features of the rat are characteristic of all mammals?

2. What sensory organs are present on the head?

MUSCULAR SYSTEM

Muscles are designed with one basic purpose in mind—movement. Muscles work to either move an animal through its environment or move substances through an animal. In vertebrates, there are three basic types of muscle tissue: (1) skeletal muscle and (2) cardiac muscle, both of which possess striated fibers, and (3) smooth muscle, sometimes called visceral muscle. Some of these muscles, like skeletal muscle, can be voluntarily controlled by the animal, while others, like cardiac and many smooth muscles, produce involuntary actions that are regulated by the autonomic nervous system. The muscles that you will dissect will be the skeletal muscles associated with the thoracic and abdominal regions of the body.

1. To observe the thoracic and abdominal musculature of the rat, you must first remove the skin from these regions. Place the rat on its dorsal surface in your dissecting pan and make a small incision through the skin (but **not** through the body wall!) with scissors along the midventral line from the tip of the jaw caudally toward the tail.

2. Separate the skin from the underlying muscles carefully, using a blunt probe where necessary.

3. Make two transverse incisions through the skin around the body, one around the jaw and the other just in front of the hindlimbs.

4. Make incisions in the skin around each wrist and longitudinal incisions along the ventral surface of

each forelimb toward the midventral incision made earlier.

5. Use a blunt probe to carefully tease the skin away from the muscles in the thoracic and abdominal areas. In most cases you should be able to "peel" the skin away with your fingers. Continue until you have completely exposed the muscles of your specimen.

6. After the skin is removed, you may need to clean away subcutaneous fat and membranous fascia that cover the muscles in this region of the body.

7. Use Table 18.2 and Figure 18.2 to identify the selected muscles and their actions in the rat.

TABLE 18.2 • Superficial Musculature of the Thoracic and Abdominal Regions in the Rat

Muscle Name	Action
Sternomastoideus	Turns head
Sternohyoideus	Pulls tongue backward
Acromiodeltoideus	Pulls humerus forward
Biceps brachii	Flexes forelimb
Triceps (medial head)	Extends forearm
Pectoralis major	Adducts forelimb
Pectoralis minor	Adducts forelimb
Latissimus dorsi	Pulls humerus backward
Rectus abdominis	Compresses abdomen
External oblique	Compresses abdomen and flexes trunk

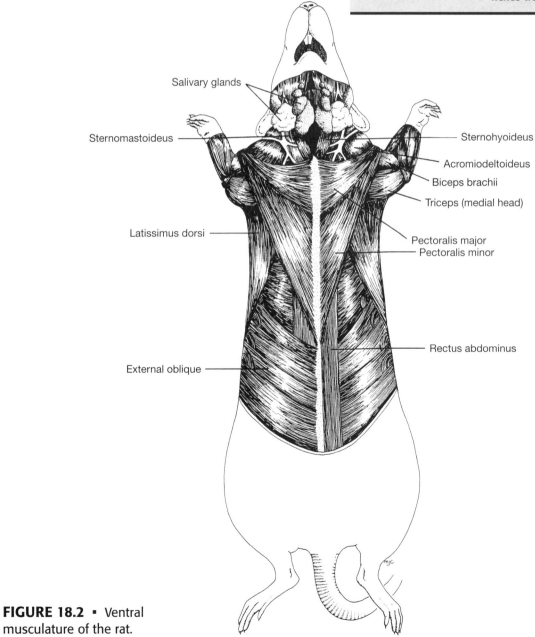

FIGURE 18.2 • Ventral musculature of the rat.

DIGESTIVE SYSTEM

The digestive system is responsible for mechanically and chemically breaking down food into smaller, usable compounds and then transporting those nutrients into the bloodstream for delivery to the individual cells of the body. This process provides the crucial raw materials and energy for all metabolic processes carried out by mammals. The extreme specialization of individual digestive organs and the efficiency of the digestive process permit mammals to sustain high metabolic rates and maintain an endothermic balance without the need for constant consumption of food.

1. Place your rat on its dorsal surface in your dissecting pan.

2. Using scissors make a shallow, longitudinal incision along the midventral line through the body wall (and rib cage) from the tip of the jaw caudally toward the anus. Orient your incision to one side of the reproductive organs to avoid damaging them.

3. Make two transverse incisions along the sides of the body just caudal to the ribs and just cranial to the hindlimbs.

4. Spread apart the ribs and the flaps of tissue in the abdominal region and pin them to the dissecting tray to expose the internal organs.

5. To obtain a clear view of the trachea and esophagus, use a blunt probe to push aside the musculature and salivary glands in the neck region.

6. Use Figure 18.3 and Table 18.3 to assist you in identifying the digestive and respiratory organs and their corresponding functions in the body.

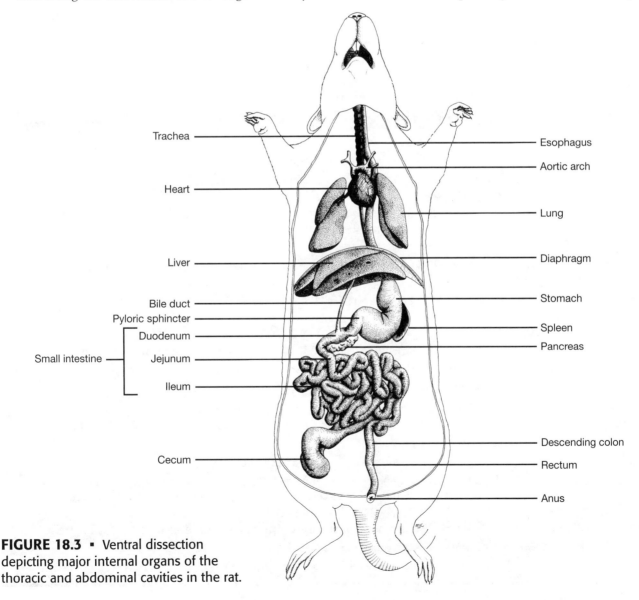

FIGURE 18.3 ▪ Ventral dissection depicting major internal organs of the thoracic and abdominal cavities in the rat.

The digestive system of the rat follows the basic mammalian blueprint with only a few exceptions. Digestion begins in the **mouth** where the **teeth** mechanically grind the food as it mixes with secretions produced by the **salivary glands**. This softened mixture passes down the **esophagus** to the **stomach** where chemical secretions from the stomach lining further the digestive process. Rats lack a gallbladder; thus bile from the **liver** is released directly into the **duodenum**, the first portion of the small intestine, through the **bile duct**. The **pancreas** produces digestive enzymes that also enter the duodenum in this region. Food leaves the duodenum and enters the second portion of the small intestine, the **jejunum**. The terminal third of the small intestine is called the **ileum**. Together these three regions of the small intestine perform the majority of nutrient absorption in mammals. At the juncture of the ileum and the colon, a large, blind-ended pouch extends into the coelomic cavity. This thin-walled sac is the **cecum** which contains anaerobic bacteria responsible for the fermentation of cellulose and other plant materials.

CAUTION: Do not puncture or open the cecum! It contains foul-smelling gases that would be extremely unpleasant to you and your fellow students.

In rats and other herbivores, the cecum is large and well-developed to handle the considerable amount of cellulose in their diets. For example, the koala obtains virtually all of its nutrition from eucalyptus leaves and has an extraordinarily long cecum (2 meters!) for such a relatively small body size. In carnivorous and omnivorous mammals, the cecum is greatly reduced in size and has a limited function in this capacity. Food enters the **colon** (i.e., large intestine) where reabsorption of water and electrolytes occurs and relatively dry feces are produced and transported to the rectum via peristalsis. The **rectum** is the final site of water reabsorption and feces dehydration and the **anus** regulates the egestion of the feces from the body.

TABLE 18.3 ▪ Internal Organs in the Rat and Their Functions

Organ/Structure	Function
Esophagus	Transports food to stomach
Trachea	Conducts air to and from lungs during respiration
Aortic arch	Transports oxygenated blood from the left ventricle of the heart to the dorsal aorta for distribution to the regions of the body
Heart	Four-chambered organ (two atria, two ventricles) that receives oxygenated blood from the lungs and pumps it via the arteries throughout the body
Lungs	Paired, multi-lobed, highly-vascularized organs for respiration
Diaphragm	Flat, muscular sheet separating thoracic and abdominal cavities used to ventilate lungs through negative pressure
Liver	Produces bile, converts glucose to glycogen for storage, detoxifies many constituents of the absorbed digested compounds
Bile duct	Transports bile from the lobes of the liver directly to the duodenum
Stomach	Produces hydrochloric acid and pepsinogen that aid in the chemical breakdown of food
Pyloric sphincter	Muscular band which regulates the flow of chyme from the stomach into the duodenum
Spleen	Stores blood, recycles worn-out red blood cells, produces lymphocytes
Pancreas	Produces digestive enzymes and delivers them through pancreatic duct to duodenum
Duodenum	Receives chyme from the stomach along with secretory enzymes from the gallbladder and pancreas
Jejunum	Responsible for majority of nutrient absorption and reabsorption of water
Ileum	Continues process of nutrient absorption and reabsorption of water
Cecum	Large, thin-walled pouch demarcating beginning of the large intestine that contains anaerobic bacteria responsible for fermentation of cellulose and other plant materials; has a reduced appearance and function in carnivores and omnivores
Descending colon	Responsible for reabsorption of water and electrolytes; transports feces to rectum via peristalsis
Rectum	Final site of water reabsorption and feces dehydration
Anus	Regulates egestion of undigested food (feces) from the body

Check Your Progress

1. Which structures/organs in the mammal are responsible for the mechanical digestion of food?

2. Which structures/organs are responsible for chemical digestion?

3. Which structures/organs are responsible for nutrient or water absorption?

4. What role does the cecum play in the digestive process of herbivores?

RESPIRATORY SYSTEM

The respiratory system of mammals is responsible for bringing a fresh supply of oxygen to the bloodstream and carrying off excess carbon dioxide. The anatomy of the respiratory tract is designed to humidify and warm the air while filtering out dust particles and germs. The lining of the nasal epithelium is covered with fine hairs that capture these foreign particles and prevent them from passing into the lungs where they may infect the body. Similarly, as air is exhaled it is cooled and dried, thus reducing the amount of heat and moisture that terrestrial mammals lose through respiration.

1. Locate the long **trachea** descending from the oral region to the **lungs**. Notice the cartilaginous rings present to keep the trachea from collapsing under the negative pressures generated during respiration (Fig. 18.3).

2. The trachea branches into two primary bronchi that enter the right and left lungs.

3. Identify the thin, muscular **diaphragm** lying on the cranial margin of the liver. The diaphragm is a uniquely mammalian characteristic that improves the efficiency of the respiratory process and helps mammals maintain a high metabolic rate.

Check Your Progress

1. How many distinct lobes are present on the right lung? the left lung?

CIRCULATORY SYSTEM

The circulatory (or cardiovascular) system is responsible for transporting nutrients, gases, hormones and metabolic wastes to and from the individual cells of an animal. Mammals are far too large for all of their individual cells to exchange nutrients, wastes and gases with the external world by simple diffusion. Most cells are buried too deep inside the body to effectively accomplish this task. Thus some system must be in place to efficiently exchange these products between the outside world and every cell in the organism's body. For this reason, the circulatory system is a highly-branched network of vessels that spreads throughout the entire organism.

In general the circulatory system represents a series of vessels that diverge from the heart (arteries) to supply blood to the tissues and a confluence of vessels draining blood from the tissues (veins) and returning it to the heart. Despite the extensive network of arteries and veins throughout the body, no actual exchange of water, nutrients, wastes or gases occurs in arteries or veins; their walls are too thick to permit diffusion. Extensive networks of capillary beds connecting branches of arteries and veins exist throughout the body to transfer these dissolved substances between the bloodstream and the tissues.

To simplify identification of the numerous arteries and veins, there are two general principles you should remember: (1) arteries and veins tend to be paired, especially when the organs they supply or drain are paired, and (2) a continuous vessel often undergoes several name changes along its length as it passes through different regions. Therefore to successfully identify arteries and veins it is necessary to trace them along their entire length (typically from the heart outward).

1. If you have not exposed the interior of the thoracic cavity by opening the rib cage, do so at this time.

2. The heart will be encased in a thin, pericardial membrane which should be removed carefully to properly view the heart.

3. Using a teasing needle and forceps, dissect the muscle tissue and fatty tissue away from the major arteries and veins in the neck, thoracic, abdominal and pelvic regions. This is a very tedious process and will take some time. Use Figure 18.4 as a guide.

4. If your specimen has been double-injected with latex, the arteries will appear red and the veins will appear blue. If your rat has not been injected with latex, the arteries will appear whiter and stiffer than the thin, collapsed veins. Remember that arteries are more heavily walled than veins (to accommodate higher blood pressures) and generally will be thicker and thus, more evident during dissection.

5. You may find it helpful to remove veins from the thoracic region to better view the arteries in this area. If so, only remove veins that you have identified and be careful not to damage arteries in the process. Since many veins lie adjacent to neighboring arteries, you will need to exercise caution when removing veins.

6. Use Figure 18.4 and Table 18.4 to identify the major arteries and veins in the rat.

The **spleen** is a vascular, ductless organ that plays a critical role in the circulatory system of vertebrates (Fig. 18.3). Since mammalian red blood cells do not contain nuclei, they cannot undergo cell division and thus have a finite life span. New red blood cells are continuously produced in the bone marrow and delivered to the spleen for storage. The spleen stores these cells along with excess blood and releases these products into the bloodstream as needed. Through this mechanism the spleen regulates the body's total blood volume and the relative concentration of red blood cells. The spleen also manufactures white blood cells (lymphocytes) to fend off diseases and destroys and recycles worn-out blood cells.

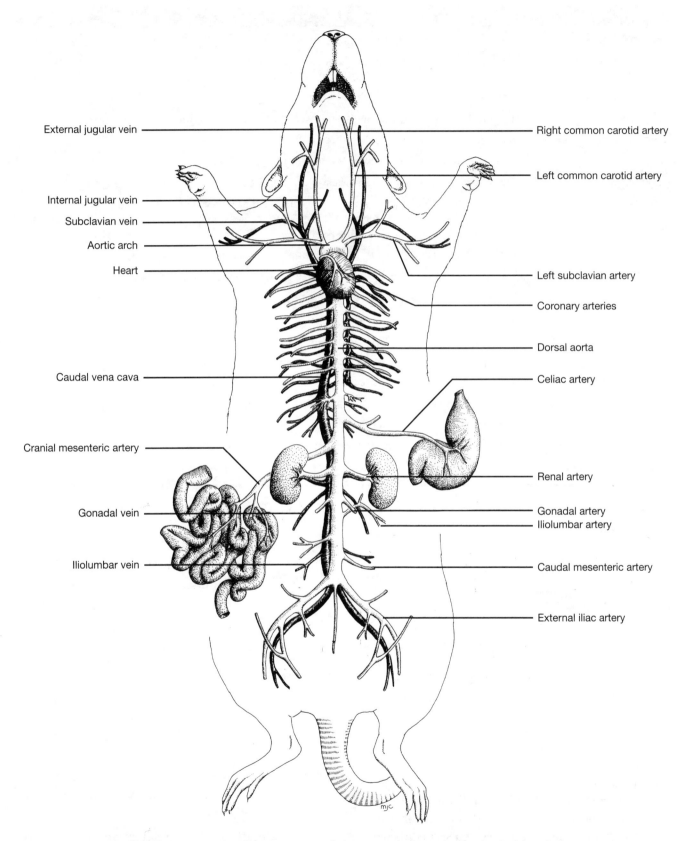

FIGURE 18.4 ▪ Circulatory system of the rat. Veins are darkly shaded for clarity.

TABLE 18.4 ▪ Major Blood Vessels in the Rat and Their Functions

Vessel Name	Function
External jugular veins	Transport blood away from the head region
Right and left common carotid arteries	Supply oxygenated blood to the head and neck regions
Internal jugular veins	Transport blood away from the neck region
Subclavian veins	Transport blood away from the lateral portions of the thoracic cavity and forelimbs
Left subclavian artery	Supplies oxygenated blood to the forelimb
Aortic arch	Supplies oxygenated blood from the heart to the entire body
Coronary arteries	Supply oxygenated blood to the musculature of the heart
Heart	Four-chambered organ (two atria, two ventricles) that receives deoxygenated blood from all regions of the body, pumps it to the lungs, then receives oxygenated blood from the lungs and pumps it to all regions of the body
Dorsal aorta	Supplies oxygenated blood to all posterior parts of the body
Caudal vena cava	Transports blood back to the heart from all posterior parts of the body
Celiac artery	Supplies oxygenated blood to the stomach, pancreas and spleen
Renal arteries	Supply oxygenated blood to the kidneys
Cranial mesenteric artery	Supplies oxygenated blood to the jejunum, ileum and colon
Gonadal arteries	Supply oxygenated blood to the testes (male) and ovaries (female)
Gonadal veins	Transport blood away from the testes (male) and ovaries (female)
Iliolumbar arteries	Supply oxygenated blood to the dorsal surface of the pelvic region
Iliolumbar veins	Transport blood away from the dorsal surface of the pelvic region
Caudal mesenteric artery	Supplies oxygenated blood to the mesentery of the large intestine
External iliac arteries	Supply oxygenated blood to the hindlimbs

UROGENITAL SYSTEM

The urogenital system is comprised of both excretory and reproductive organs. Excretory organs are responsible for eliminating metabolic wastes that the body produces from cellular respiration and for maintaining a homeostatic balance between the levels of fluids, electrolytes, sugars, hormones and proteins in the body. Remember, excretion is an entirely different process from that which expels indigestible products through the anus! Excretion and egestion (or defecation) are different processes, handled by completely different systems in mammals.

Reproductive organs are responsible for producing the gametes that will ultimately fuse with the corresponding gametes of the opposite sex. In addition to reproduction, the testes and ovaries produce many of the hormones associated with the development and maturation of primary and secondary sexual characteristics and which drive the repertoire of sexual behaviors indicative of most mammals.

1. Using a teasing needle, carefully dissect away the membranous tissue surrounding one of the kidneys. Take care not to destroy the adrenal gland which sits along the cranial margin of the kidney (probably imbedded in fat) or any of the ducts and blood vessels in the area.

2. If your specimen is a male, be careful not to damage the vas deferens which "loops" around the ureter.

3. Clean the area around the **kidney** to expose the **renal blood vessels** and the **ureter** passing from the medial margin of the kidney caudally toward the **urinary bladder**.

4. Use Table 18.5 and Figures 18.5 and 18.6 to identify the major excretory organs and their functions in the rat.

5. To completely uncover all of the reproductive structures you must cut longitudinally through the pubic symphysis with a scalpel. *Cut carefully*

and start your incision to one side of the median plane of the pelvis to avoid actually cutting through underlying structures. It is preferable to only partially cut through the symphysis and then apply downward (lateral) pressure to each of the hindlimbs to complete the separation.

6. Regardless of the sex of your specimen, you are expected to be familiar with the reproductive structures of both male and female rats, so work closely with another group that has a specimen of the opposite sex.

7. Use Table 18.5 and Figures 18.5 and 18.6 to identify the reproductive organs and their functions in the rat.

The **scrotum** of the male houses the **testes**, small bean-shaped structures which are the site of sperm production. Your first task in dissecting the male reproductive system is to locate the spermatic cords which leave the scrotum and enter the abdominal wall. Each testis is enclosed within a cremasteric pouch. At the cranial end of the cremasteric pouch, a narrow tube should be evident. This is the **spermatic cord** which contains the vas deferens, the spermatic artery and vein, lymphatic vessels and numerous nerves. The testes will be located within the cremasteric pouch.

8. Carefully make a slit in the cremasteric pouch and peel it open, using scissors if necessary. Leave the testis attached to the spermatic cord, but

TABLE 18.5 ▪ Male and Female Urogenital Organs in the Rat and Their Functions

Corresponding homologous structures in the two sexes are placed in the same row.

Male Structure	Function	Female Structure	Function
Adrenal glands	Produce hormones which regulate blood pressure and metabolism	Adrenal glands	Produce hormones which regulate blood pressure and metabolism
Kidneys	Filter blood (creating urine) and responsible for osmoregulation	Kidneys	Filter blood (creating urine) and responsible for osmoregulation
Seminal vesicles	Contribute seminal fluid with nutrients for sperm, and hormones to initiate uterine contractions		
Ureters	Transport urine from kidneys to bladder for storage	Ureters	Transport urine from kidneys to bladder for storage
Urinary bladder	Stores urine prior to excretion	Urinary bladder	Stores urine prior to excretion
Vasa deferentia*	Transport sperm to urethra	Uterine horns* (and body of uterus)	Site of implantation and embryonic development
Prostate gland	Contributes seminal fluid that may aid in neutralization of acidity of vagina	Vagina	Receives penis during copulation; serves as part of birth canal
Testes	Produce sperm	Ovaries	Produce eggs
Epididymides*	Store sperm	Oviducts*	Receive eggs at ovulation; site of fertilization
Urethra	Receives seminal secretions from accessory glands	Urethra	Drains excretory products from urinary bladder (*no reproductive function*)
Cowper's glands	Contribute seminal fluid that aids in neutralization of acidity of vagina		
Penis	Deposits semen in female reproductive tract	Clitoris	Plays a role in sexual sensation and stimulation
		External vaginal orifice (homologous to distal portion of urethra in male)	Common chamber for the release of urinary products and acquisition of sperm

* These structures technically are not homologous since they develop from separate embryonic tubes.

separate it from the tissue of the cremasteric pouch.

Cupped around the side of each testis is a highly-coiled system of tubules known as the **epididymis** (Fig. 18.5). Sperm are produced within the **seminiferous tubules** of the testis and are stored along the length of the epididymis. Upon ejaculation, sperm leave the epididymis and travel through the **vas deferens** toward the **urethra**. Trace along the length of the spermatic cord to see the path sperm travel as they move out of the epididymis through the vas deferens (which loops around the ureter) toward the prostate region. Notice that there is an opening in the abdominal wall (the **inguinal canal**) through which the spermatic cord passes from the scrotum into the abdominal cavity. The **penis** is enclosed in a sheath and held along the ventral wall of the abdomen.

Careful dissection of this area should reveal several accessory glands. Together, the **prostate gland, seminal vesicles** and **Cowper's glands** contribute fluid to the sperm, accounting for over 60% of the total volume of the semen. This fluid is thick and contains mucus (to prevent the sperm from drying out) and large amounts of fructose (to provide energy for the sperm). In addition, semen is highly basic to neutralize the acidic environment of the vagina and increase the chances of survival for the sperm.

The paired female gonads are called **ovaries** (Fig. 18.6). They are located in the abdominal region caudal to the kidneys and can be identified by their small, round appearance. The ovaries are anchored to the dorsal wall of the abdominal cavity and to each uterine horn by small ligaments. Attached to each ovary is a coiled oviduct. The **oviduct** receives the mature oocyte (egg) when it is released from the ovary at the time of ovulation.

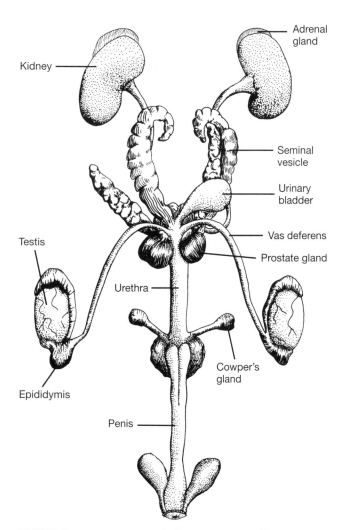

FIGURE 18.5 ▪ Urogenital system of male rat.

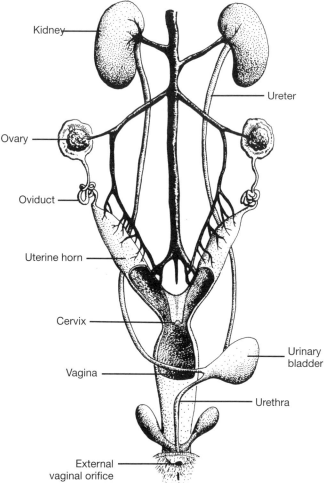

FIGURE 18.6 ▪ Urogenital system of female rat.

Despite the close association, there is no actual physical connection between the oviductal opening and the ovary. Small finger-like projections of the oviduct generate movements that sweep the egg into the ostium of the oviduct. The epithelial lining of the oviduct is ciliated and creates a current that propels the egg along the length of the oviduct toward the horn of the uterus. Fertilization typically occurs in the oviduct, but implantation of the embryos occurs further along in the uterus. In rats, the uterus is divided into two **uterine horns** in which embryonic development of the fetuses occurs. The two uterine horns converge on the **cervix**, a cartilaginous constriction marking the juncture with the **vagina**.

In humans, the uterine horns (known as fallopian tubes) are substantially reduced, since implantation and embryonic development occur in the body of the uterus. Due to larger litter sizes, rats require a much greater area for young to develop, and the extensive size of the two uterine horns accommodates this need.

 Questions For Review

1. Number the following organs in the order that represents the correct path of food through the digestive tract of a mammal:

    _____ anus   _____ mouth   _____ rectum

    _____ ileum   _____ jejunum   _____ colon

    _____ stomach   _____ esophagus   _____ cecum

    _____ duodenum

2. What is the adaptive value of an extremely long digestive tract?

3. What is the primary evolutionary reason for mammals to develop and house their lungs inside the body cavity?

4. Explain the functional relationship between the respiratory system and the circulatory system.

5. The mammalian heart has _____ chambers.

6. Differentiate between excretion and egestion.

7. In the mammalian excretory system, urine is produced in the kidney and travels through the

   _____ to the _____ where it is temporarily stored.

8. Match the structure on the left with its *most appropriate* function on the right.

   _____ epididymis   a. produces eggs

   _____ ovary   b. releases urine to outside environment

   _____ oviduct   c. captures recently "erupted" egg

   _____ testis   d. site of embryo implantation/development

   _____ uterine horn   e. produces sperm

   _____ vas deferens   f. transports sperm to penis during ejaculation

                                 g. stores sperm

                                 h. contributes extra fluid to semen

                                 i. cartilaginous constriction of cranial end of vagina

References

Abramoff, P. and R. G. Thomson. 1994. *Laboratory Outlines in Biology VI*. W. H. Freeman and Company: New York, New York.

Berger, A. J. 1960. The musculature. Pp. 301-344 *in*: Biology and Comparative *Physiology of Birds*, vol. 1 (A. J. Marshall, editor). Academic Press: New York, New York, 518 pp.

Brooke, M. and T. Birkhead. 1991. *The Cambridge Encyclopedia of Ornithology*. Cambridge University Press: Cambridge, Massachusetts.

Brusca, R. C. and G. J. Brusca. 1990. *Invertebrates*. Sinauer Associates: Sunderland, Massachusetts.

Campbell, N. A., J. B. Reece and L. G. Mitchell. 1999. *Biology* (5th ed.). Benjamin/Cummings: Menlo Park, California.

Chiasson, R. B. 1972. *Laboratory Anatomy of the Pigeon* (2nd ed.). Wm. C. Brown: Dubuque, Iowa.

Dolphin, W. D. 1999. *Biological Investigations: Form, Function, Diversity and Process* (5th ed.). WCB/McGraw-Hill: Boston, Massachusetts.

Dorit, R. L., W. F. Walker, and R. D. Barnes. 1991. *Zoology*. Saunders: Philadelphia, Pennsylvania.

Drewes, C. 1996. "Those wonderful worms." *Carolina Tips*. 59(3):17-20.

Drewes, C. and K. Cain. 1999. As the worm turns: locomotion in a freshwater oligochaete worm. *Am. Bio. Teacher*. 61(3):438–442.

Glase, J. C., M. C. Zimmerman, and J. A. Waldvogel. 1992. Investigations in orientation behavior. Pp. 1-16 *in*: *Tested Studies for Laboratory Teaching*, vol. 6 (C. A. Goldman, S. E. Andrews, P. L. Hauta, and R. Ketcham, editors). Proceedings of the 6th Workshop/Conference of the Association for Biology Laboratory Education (ABLE), 161 pp.

Gunstream, S. E. 1996. *Explorations in Basic Biology* (7th ed.). Prentice Hall: Upper Saddle River, New Jersey.

Halliday, T. R. and K. Adler. 1987. *The Encyclopedia of Reptiles and Amphibians*. Facts on File Inc.: New York, New York.

Hickman, C. P., F. M. Hickman and L. B. Kats. 2001. *Laboratory Studies in Integrated Principles of Zoology* (10th ed.). McGraw-Hill: Boston, Massachusetts.

Hickman, C. P. Jr., L. S. Roberts, and F. M. Hickman. 1990. *Biology of Animals* (5th ed.). Times Mirror/Mosby: St. Louis, Missouri.

Hopkins, P. M. and D. G. Smith. 1997. *Introduction to Zoology: A Laboratory Manual* (3rd ed.). Morton: Englewood, Colorado.

Lasiewski, R. C. 1972. Respiratory function in birds. Pp. 287-342 *in*: *Avian Biology*, vol. 2 (D. S. Farner, J. R. King, and K. C. Parkes, editors). Academic Press: New York, New York. 612 pp.

Lytle, C. F. 1996. *General Zoology Laboratory Guide* (12th ed.). Wm. C. Brown Publishers: Dubuque, Iowa.

Mader, D. R. 1995. Reptilian anatomy: discover exactly what's inside your reptiles, and where. *Reptiles*. 3(2):84-93.

Miller, S. A. 1994. *General Zoology Laboratory Manual* (3rd ed.). Wm. C. Brown: Dubuque, Iowa.

Mitchell, L. G., J. A. Mutchmor, and W. D. Dolphin. 1988. *Zoology*. Benjamin/Cummings: Menlo Park, California.

Nickel, R., A. Schummer, E. Seiferle, W. G. Siller, and P. A. L. Wight. 1977. *Anatomy of the Domestic Birds*. Springer-Verlag: New York, New York.

Pechenik, J. A. 1996. *Biology of the Invertebrates* (3rd ed.). Wm. C. Brown Publishers: Dubuque, Iowa.

Perry, J. W., D. Morton and J. B. Perry. 2002. *Laboratory Manual for Starr and Taggart's Biology: The Unity and Diversity of Life and Starr's Biology: Concepts and Applications*. Brooks/Cole: Pacific Grove, California.

Purves, W. K., D. Sadava, G. H. Orians and H. C. Heller. 2001. *Life: The Science of Biology* (6th ed.). Sinauer Associates: Sunderland, Massachusetts.

Ruppert, E. E. and R. D. Barnes. 1994. *Invertebrate Zoology* (6th ed.). Saunders College Publishing: Fort Worth, Texas.

Tyler, M. S. 1994. *Developmental Biology—A Guide for Experimental Study*. Sinauer: Sunderland, Massachusetts.

Van De Graaff, K. M. and J. L. Crawley. 2001. *A Photographic Atlas for the Biology Laboratory* (4th ed.). Morton Publishing: Englewood, Colorado.

Van De Graaff, K. M. and J. L. Crawley. 1998. *A Photographic Atlas for the Zoology Laboratory* (3rd ed.). Morton Publishing: Englewood, Colorado.

Vodopich, D. S. and R. Moore. 2002. *Biology Laboratory Manual* (6th ed.). McGraw-Hill: Boston, Massachusetts.

Wachtmeister, H. F. E. and L. J. Scott. 1997. *Encounters with Life: General Biology Laboratory Manual* (5th ed.). Morton Publishing: Englewood, Colorado.

Walker, W. F. Jr. 1998. *Dissection of the Frog* (2nd ed.). W. H. Freeman and Company: New York, New York.

Walker, W. F. Jr. and D. G. Homberger. 1997. *Anatomy and Dissection of the Rat* (3rd ed.). W. H. Freeman and Company: New York, New York.

Wischnitzer, S. 1993. *Atlas and Dissection Guide for Comparative Anatomy* (5th ed.). W. H. Freeman and Company: New York, New York.

-A-

abdomen — (1) the posterior section of the body in arthropods; (2) the body cavity caudal to or below the thoracic cavity containing the viscera

abduct — movement away from the midline of the body

aboral — referring to the side or surface of the body furthest from the mouth

acoelomate — animal whose central space is filled with tissue (mesoderm) and in which no true body cavity exists

acontia — "internal tentacles" that contain cnidocytes and aid in subduing live prey taken into the gastrovascular cavity of cnidarians

actin — thin, proteinaceous filaments that overlap and alternate with myosin filaments to generate muscular contractions

action — the influence a muscle has on the motion of a joint or appendage

adaptive radiation — the evolution of numerous species from a common ancestor following migration into a new environment

adduct — movement toward the midline of the body

adipose tissue — type of connective tissue that stores or sequesters food for the body in the form of fat droplets

agar — a gelatinous extract of red algae used as a stabilizing agent in culture media and some foods

air-filled bladder — swollen, spherical gas-filled membrane that allows the photosynthetic regions of some algae species to float near the water surface

air sacs — thin, membranous pouches located throughout visceral cavity in birds; used in respiration

alginic acid — an insoluble, colloidal extract of brown algae used as an emulsifier in food and cosmetic products

Alveolata — monophyletic candidate kingdom of unicellular protists whose members possess small cavities under their cell surfaces

alveoli — small cavities found beneath the surface of protists in the candidate kingdom Alveolata

ambulacral grooves — indentations along the arms of sea stars that contain the tube feet

amniotic egg — egg (of reptile, bird or mammal) surrounded by a fluid-filled sac that encloses the embryo and protects it from desiccation

amoebocytes — mobile cells specialized for distributing food throughout the sponge and for producing its skeleton

animal pole — region of an egg in which an embryo develops and which appears darker in coloration in comparison to the vegetal pole

antagonistic — working in opposition to or limiting the movement of a paired structure

antennae — moveable, segmented organs of sensation located on the heads of arthropods

Apicomplexa — group of unicellular, parasitic protists characterized by an apical complex of organelles used to penetrate the host cell

apopyle — opening through which water passes out of the radial canals into the spongocoel in sponges

archenteron — (syn. = gastrocoel) hollow tube that is created by the arrangement of invaginated cells in the gastrula; embryonic precursor to the gut

Aristotle's lantern — structure in sea urchins that anchors teeth and coordinates their movements for scraping and chewing

arm — supports microscope body, stage and adjustment knobs

arteries — vessels which conduct blood away from the heart

asexual reproduction — type of reproduction that occurs without genetic recombination and results in the production of offspring that are genetically identical to the parent; may occur by simple fission of a single cell, budding or fragmentation

asters — array of microtubule filaments by which the centriole pairs anchor themselves to the cell membrane at the conclusion of prophase

asymmetry — lack of symmetry; irregular arrangement of body parts with no plane of symmetry to divide them into similar halves

atriopore — cephalochordate structure that discharges water from the atrium to the external environment

atrium — (1) internal chamber of the heart which receives blood from the body and channels it to a ventricle; (2) in cephalochordates, the ventral body chamber that receives water passed from pharynx through gill slits

auricles — (1) primitive chemoreceptors found in some invertebrate phyla (i.e., Platyhelminthes); (2) flap-like outer regions covering the cranial portions of the atria in many vertebrates' hearts

autotrophic — characteristic of an organism that produces its own organic food molecules, usually through photosynthesis

axon — elongated portion of a neuron that transmits electrical impulses away from the cell body

-B-

basal disc — terminal portion of stalk in cnidarian polyps that serves as the primary point of attachment to the substrate

base — part of microscope that supports microscope unit

behavior — any overt response by an animal to a stimulus

bilateral symmetry — body parts are divided into similar halves (mirror images) by a single plane of symmetry

bile duct — canal that transports secretions from the liver (and from the pancreas in some animals) to the duodenum or stomach

binary fission — type of cell division by which many unicellular protists and bacteria reproduce asexually; generally involves a simple mitotic division of the parent cell producing two genetically identical daughter cells

bipinnaria larva — immature developmental stage of sea stars immediately following the gastrula stage; characterized by bilateral symmetry and the beginning of organogenesis

biramous — having two branches

blade — broad, flat, leafy portion of an algae, modified to absorb light

blastocoel — hollow, fluid-filled cavity within the blastula

blastomeres — individual cells of a blastula

blastopore — the opening to the outside of the embryo marking the site of the invagination of cells toward the center of the blastocoel

blastula — embryonic stage of development containing 64+ cells and characterized by a migration of cells toward the periphery of the embryonic sphere creating a hollow cavity deep within the spheroid of cells

blood — fluid connective tissue that typically contains respiratory pigments and serves as a medium for nutrient and gas exchange between tissues and the external environment

body — microscope housing that keeps ocular and objective lenses in proper alignment

bone – hard, calcified connective tissue used for structural support

brain – part of the central nervous system responsible for processing and integrating nerve impulses gathered from all sensory organs and receptors and for initiating motor impulses

branchial hearts – smaller, muscularized chambers in cephalopods that receive deoxygenated blood from all parts of the body and pump blood to the gills

bronchi – (sing. = bronchus) major divisions of the trachea through which oxygen is supplied to and carbon dioxide is removed from the lobes of the lungs

bud – asexual outgrowth of the parent that pinches off when mature and lives independently; common in cnidarians

-C-

canaliculi – tiny, fingerlike projections in the lamellae of bone tissue through which nutrients are transported to the osteocytes

carapace – hard, bony, dorsal portion of the shell of turtles

cardiac chamber – first stomach region in some arthropods and echinoderms in which mechanical digestion of food occurs

cardiac muscle – striated muscle found in the walls of the heart

carrageenan – extract of red algae used as a stabilizer in paints and cosmetics and as an emulsifying agent in many foods

caudal – referring to a point of reference toward the tail

cecum – blind projection located at the junction of the ileum and colon that serves as a sac for fermentation of cellulose; plays a prominent role in the digestive process of most herbivores, but is reduced in omnivores and carnivores

cell body – portion of a neuron that contains the nucleus and other organelles

cell cycle – the events that encompass the entire lifecycle of a cell from one division to the next

cell differentiation – process by which newly formed embryonic cells take on specialized functions and undergo shifts in morphology to accommodate these new tasks

cell division – formation of new cells through mitosis or meiosis

cell migration – the organization and movement of groups of cells to create an animal's body shape

cell wall – protective, rigid outer layer of the cells of plants, fungi, bacteria and many protists; may consist of cellulose, calcium carbonate, silica or other materials

cellular slime mold – macroscopic protist consisting of large numbers of amoeboid cells containing single haploid nuclei

cellulose – a structural polysaccharide of cell walls, consisting of glucose monomers joined together

cellulose plates – structural element of dinoflagellates found beneath the outer plasma membrane providing support and shape to their cells

centric – describes diatoms that are circular, oval or elliptical in shape and have radial symmetry

centromere – centralized region joining adjacent sister chromatids

cephalization – concentration of nervous tissue and sensory structures at the cranial end of the body

cephalothorax – fused portion of the arthropod body consisting of the head and thorax regions

cervix – cartilaginous constriction demarcating the cranial boundary of the vagina

chlorophyll – green pigment located within plastids in autotrophic protists and plants that functions in the light reactions of photosynthesis

chloroplast – organelle found in autotrophic protists and plants that absorbs sunlight and drives photosynthetic reactions to produce sugars

choanocytes – specialized cells lining the interior surface of the radial canals in sponges which trap and engulf small food particles

chondrocytes – cartilage-producing cells

Chrysophyta – taxonomic group containing diatoms – unicellular autotrophs possessing two-part cell walls made of silica

cilia – short cellular appendages specialized for locomotion; found in many protists

Ciliophora – taxonomic group containing unicellular, heterotrophs possessing cilia and two types of nuclei: macronuclei and micronuclei

circular canal – circular extension of the gastric pouches that distributes nutrients to outer rim of jellyfish

cleavage furrow – shallow indentation formed along cell margin of the metaphase plate in a dividing cell

cleavage – cell division

clitellum – unsegmented band of tissue found in some groups of annelids; used in reproduction

cloaca – common chamber for the elimination or release of digestive, excretory and reproductive products; in reptiles and birds the cloaca is partitioned into three chambers: the urodeum, the coprodeum and the proctodeum

closed circulatory system – cardiovascular system in which blood is confined to and travels through a system of arteries, veins and capillaries

cnidocytes – stinging cells on cnidarian tentacles used for defense and for capturing food

coarse focus adjustment – control which moves microscope stage up or down to focus image

coelom – body cavity lined on both sides by mesoderm

coenosarc – common gastrovascular cavity shared among colonial cnidarian polyps that transports nutrients throughout the colony

collar – fleshy border separating head-foot from visceral mass (mantle) in some cephalopods

complete digestive tract – digestive tract that passes food through in one direction from the mouth to the anus

condenser adjustment – control which moves microscope condenser lens up or down to focus light

condenser – lens mounted beneath microscope stage that focuses the light beam on the specimen

conjugation – process that occurs in some protists and bacteria in which genetic material from one cell is transferred to another cell, or genetic material from two cells is simultaneously transferred

conjugation tube – cytoplasmic bridge that forms between adjacent cells during conjugation in some algal species to permit genetic material from one cell to move across to the other cell

connective tissues – groups of cells that work together to bind, support and protect body parts and systems

contractile stalk – thin extension of the plasma membrane of certain protists that they use to extend and retract their cell bodies in relation to their substrate

contractile vacuole – spherical organelle designed to pump excess water out of cells

convergent evolution – the independent development of similarity between species as a result of similar selection pressures typically generated by comparable ecological roles

coprodeum – anterior-most chamber of the cloaca in reptiles and birds; receives fecal matter directly from the large intestine

copulatory organs – structures used for sexual intercourse

Cowper's glands – accessory male reproductive glands that contribute fluid to the semen

cranial – referring to a point of reference toward the head

crossing over – a process in which small sections of chromosome are exchanged between neighboring homologous chromosomes during meiosis I

cuticle – thin, epidermal secretion that covers the outer surface of the body and protects the organism from external chemical and mechanical dangers and/or desiccation

cyst – dormant, resistant stage of cells that allows them to survive prolonged periods during which environmental conditions are unfavorable for growth or survival

cytopharynx – tubular opening in ciliates through which food passes and is incorporated into food vacuoles

cytoplasmic streaming – phenomenon in which cytoplasm of a cell flows from one region of the cell body to another, often as a means of distributing food, gases or other substances

-D-

daughter colony – product of reproductive process in colonial forms of algae such as *Volvox*

dendrites – parts of a neuron which typically receive electrical impulses from neighboring neurons or sensory receptors and transmit them to the cell body

depth of field – the thickness of an image that is in focus at any point in time when looking through a microscope

dermal branchiae – echinoderm organs of gas exchange and excretion

deuterostome – animal whose mouth develops from the second embryonic opening and whose embryonic cells divide by radial cleavage

diaphragm – muscular sheet separating the thoracic and abdominal cavities; used to ventilate the lungs of mammals

diatomaceous earth – type of soil formed by the accumulation of siliceous diatom cell walls over hundreds of millions of years

digestive gland – organ that secretes digestive enzymes into stomach and intestine in many invertebrate phyla to assist in the breakdown of food

dioecious – referring to an organism that contains only male or female reproductive structures

diploblastic – referring to an animal whose body develops from only two embryonic tissue layers

direct development – form of embryonic development in which no larval stage is present; young emerge resembling "miniature" adults

disposable pipettes – devices used to aliquot small volumes of fluid; for one-time use only

distal – referring to a point of reference farther from the body's median plane or point of attachment than another structure

dorsal lip – crescent-shaped groove along the surface of the blastula which marks the opening of the blastopore

dorsal – referring to a point of reference nearer the backbone

ductus deferens – (syn. = vas deferens) tubule in male animals that transports sperm

duodenum – anterior portion of the digestive tract that receives secretions from the liver and pancreas through the common bile duct for further breakdown of food from the stomach

-E-

ecdysis – (syn. = molting) the periodic act of shedding the skin or exoskeleton

ectoderm – outermost of the three primary embryonic germ layers which gives rise to the outer covering of animals and, in some phyla, the nervous system

ectoparasite – parasite that attaches itself to the outside of the host organism to obtain its nourishment

ectotherm – an animal that derives the majority of its body heat from external sources

elastic cartilage – gelatinous connective tissue that contains fine collagen fibers and many flexible fibers; found in the ear, nose and voice box of humans

endoderm – the innermost of the three primary embryonic germ layers which gives rise to many major internal organs and linings

endoparasite – parasite that lives within the body of its host

endoskeleton – animal support system enclosed beneath the outer body surface (i.e., bones of humans, ossicles of echinoderms)

endotherm – an animal that derives the majority of its body heat from internal metabolic sources

epidermis – outer tissue covering in animals; derived from ectoderm

epididymis – highly-coiled tubule system that cups around the testis and serves as a storage unit and transportation canal for mature sperm

epithelial tissues – groups of cells which cover external surfaces for protection or line the internal surfaces of body cavities and vessels

erythrocytes – red blood cells; modified for transport of oxygen and carbon dioxide to and from the body tissues

esophagus – thin tube connecting the mouth to the stomach

eucoelomate – animal with a central body cavity that lies between layers of mesoderm

euglenoids – category of protists characterized by an anterior pocket that bears one or two flagella extending from the anterior end of the organism

Euglenozoa – monophyletic candidate kingdom of protists containing euglenoids and kinetoplastids

eukaryotic – characteristic of cells that contain a membrane-enclosed nucleus and other membrane-enclosed organelles

excretory canals – longitudinal channels running along the outer margins of the tapeworm body that expel metabolic waste products

extensor – a muscle that straightens or extends a body part

eyes – image-forming, photoreceptive organs of sight

eyespots – rudimentary photoreceptors capable of detecting shadows and extremes in light; found in some invertebrate phyla (i.e., Platyhelminthes)

-F-

fertilization membrane – thin membrane encircling zygote; produced by the egg immediately following fertilization to prevent polyspermy (fertilization of the egg by multiple sperm)

fertilization – the process of contact between sperm and egg, entry of the sperm into the egg and fusion of the egg and sperm pronuclei to form a zygote

fetal membranes – embryonic tissues involved in pregnancy; chorion, amnion and allantois.

field of view – the circular area seen when looking through a microscope

filopodia – thin protoplasmic extensions put out by clumps of aggregating sponge cells

fine focus adjustment – control which permits precise focusing of a microscope

flagellum – (pl. = flagella) long cellular appendage specialized for locomotion

flame cells – excretory units common in many flatworms that collect metabolic wastes from nearby regions and channel wastes to excretory pores located on the body surface

flexor – a muscle that bends one part of the body toward another part

food vacuole – small, spherical organelle containing enzymes to digest food

foot – extensible, muscular organ used by molluscs for locomotion

fragmentation – method of reproduction in which organism breaks into smaller pieces capable of surviving on their own

frontal plane – longitudinal section dividing an animal into dorsal and ventral parts

fruiting bodies – specialized structures of slime molds which produce spores

fucoxanthin – brown photosynthetic pigment found in many protists

-G-

gallbladder – small, membranous, muscular sac which stores bile produced by the liver

gametogenesis – the production of egg or sperm cells

gastric mill – chitinous teeth that grind food located within cardiac stomach of crustaceans

gastric pouch – one of four divisions of the gastrovascular cavity of jellyfish for digestion of food

gastrocoel – (syn. = archenteron) hollow tube created by the arrangement of invaginated cells in the gastrula; embryonic precursor of the gut

gastrodermis – inner tissue lining of the digestive system in many animals; derived from endoderm

gastrovascular cavity – (syn. = coelenteron) central digestive compartment in some invertebrates; characterized by a single opening which functions as both mouth and anus

gastrula – embryonic stage of development characterized by the invagination of the cell surface and the migration of blastomeres toward the center of the gastrocoel

genital pore – opening for the release and/or introduction of sperm or eggs

germ layers – primary embryonic tissue layers that differentiate during gastrulation

gill bars – semi-rigid structures in pharyngeal region of chordates; only functional in some groups (i.e., urochordates and cephalocordates) for trapping small food particles on mucus coating

gill filaments – highly-vascularized, feathery extensions of the respiratory structures in fishes which provide a large surface area for gas exchange underwater

gill rakers – hard, serrated structures in fish gills that protect the gill apparatus from the passage of coarse material that could damage gill surfaces

gill slits – spaces between gill bars that permit water to flow across gill bars

gills – highly vascularized, localized extensions of the body surface of aquatic animals modified for gas exchange

gizzard – thick-walled, muscular digestive pouch in birds that pulverizes and churns food prior to passage into the intestine

glial cells – specialized cells which assist in propagating nerve impulses and provide a nutritive role for neurons

gonad – reproductive organ that produces sperm or eggs

gonangium – reproductive polyp in some colonial cnidarians

gray crescent – delineating line between the animal pole and vegetal pole of amphibian eggs which forms on the opposite side of the egg from the entry point of the sperm

-H-

haploid gametes – reproductive cells that contain half the number of chromosomes as the somatic cells

Haversian canals – tiny, narrow pathways around which lamellae form layered rings in bone tissue

heart – muscularized component of circulatory system responsible for pumping blood throughout body

hemipenes – paired copulatory organs of male reptiles

hepatic cecum – lateral outpocket of intestine in cephalochordates which is responsible for intracellular digestion of small food particles and lipid and glycogen storage

hermaphroditic – referring to an animal which possesses both male and female sex organs, but is not capable of self-fertilization (i.e., earthworms)

heterotrophic – characteristic of organisms that obtain nutrients by ingesting or absorbing organic material from external sources

hinges – regions of flexion in the plastron of some turtle species

holdfast – specialized, terminal region of algae that anchors them to the substrate

holoblastic cleavage – cell division which completely separates the parent cell into roughly equal-sized, but distinct daughter cells

homeothermic – maintaining a nearly constant internal body temperature

homologous structures – features in different species that are similar due to common ancestry

hooks – sharp, pointy attachments characteristic of many cestodes; modified for attachment of the parasite to its host

hyaline cartilage – gelatinous connective tissue composed primarily of chondrin with thin collagen fibers; situated between bones where it cushions the surfaces of joints

hydranth – feeding polyp in some colonial cnidarians

hypostome – enlarged mound of tissue that surrounds the mouth in cnidarians

-I-

ileum – distal portion of the small intestine in vertebrates extending from the jejunum to the cecum; primarily responsible for absorption of nutrients

illuminator – source of light in a microscope

incurrent canals – small openings in the body surface of sponges through which water carrying oxygen and nutrients enters

inguinal canal – opening in the abdominal wall through which the spermatic cord passes

ink sac – large sac common in cephalopods that opens into the anus and secretes a dark brown or black fluid when the animal is alarmed

insertion – moveable end of a muscle attachment

intercalated discs – gap junctions between adjacent cells of cardiac muscle that permit expedient and synchronous contractions of the heart

internal fertilization – reproductive method in which sperm are deposited within the female reproductive tract and fusion of sperm and egg occurs inside the body

interphase – the period in the cell cycle when the cell is not dividing and cellular metabolic activity is high

intestine – long, sometimes coiled, digestive organ which primarily absorbs nutrients, ions and water

iris diaphragm – control mounted beneath microscope stage near condenser that regulates amount of light illuminating specimen

isogamous – characteristic of gametes from opposite sexes that appear identical in size and shape

isolecithal – classification of eggs with relatively little yolk that is evenly distributed throughout the cytoplasm

-J-

jejunum – middle portion of the small intestine in vertebrates extending from the duodenum to the ileum; primarily responsible for nutrient absorption

-K-

keel – prominent ventral protrusion of the sternum common in birds adapted for flight

keratinized scales – shingle-like, overlapping body covering characteristic of reptiles; composed of fibrous protein to produce a hard, dry outer layer

kidney – excretory organ which filters the blood to create a highly-concentrated metabolic by-product (urine); also responsible for maintaining a homeostatic balance of salts, fluids and ions within the body (osmoregulation)

kinesis – random behavior that affects the general rate of movement or degree of turning in an animal

kinetochore – site of attachment of a chromosome at its centromere to a spindle fiber during cell division

kinetoplast – characteristic organelle of kinetoplastids consisting of a single, large mitochondrion that contains extra-nuclear DNA

kinetoplastids – group of protists characterized by the presence of kinetoplasts (see previous definition)

-L-

lacunae – hollow chambers which contain chondrocytes (in cartilage) or osteocytes (in bone)

lamellae – thin, concentric layers of a hard, calcified matrix which gives bone its characteristic appearance

large intestine — distal portion of the digestive tract that completes nutrient absorption and extracts water and ions from the fecal waste

larva — free-living, sexually immature stage in certain animal lifecycles

lateral canals — short branches of the radial canals in the echinoderm water vascular system which terminate at the tube feet

lateral — referring to a point of reference farther from the median plane

leukocytes — white blood cells; typically function in immune system

lipids — family of compounds including fats, phospholipids, and steroids, that are insoluble in water

liver — relatively large organ which produces bile and other secretions to facilitate digestion of food; also plays a role in lipid and glycogen storage and regulates blood glucose levels

loose connective tissue — type of connective tissue composed of loosely scattered cells surrounded by a clear, gelatinous matrix

lung — internal organ of respiration

-M-

macronucleus — organelle of ciliates containing many copies of a few genes and primarily responsible for metabolic processes of the cell

macroscopic protists — informal designation for collection of protists that are relatively large and easily observable to the naked eye; includes multicellular algae and slime molds

madreporite — porous entrance to the water vascular system that serves as both pressure regulator and simple filter in most echinoderms

mammary glands — modified tissues on the ventral surface of mammals that secrete milk to nourish young

mammary papillae — small protuberances on the ventral surface of mammals; in adult females the papillae will develop into teats through which the mammary glands secrete milk

mantle — thin, fleshy membrane characteristic of all molluscs; secretes the shell

manubrium — stalk of fleshy tissue present in hydrozoan medusae that supports the mouth

marginal tentacles — cnidarian tentacles that provide sensory information and are used for defense and locomotion

matrix — ground substance found in connective tissue produced by the living cells which it contains; may be liquid (i.e., blood), gelatinous (i.e., cartilage) or solid (i.e., bone)

medial — referring to a point of reference closer to the median plane

median plane — the sagittal plane running along the midline of the animal

medusa buds — young, immature medusae produced within the reproductive polyps of some cnidarians

medusa — stage in the cnidarian lifecycle represented by a circular, free-swimming form resembling the familiar jellyfish in its morphology

meiosis — two-stage process of cell division in sexually reproductive organisms that results in gametes with half the chromosome number of the original cell

meniscus — the curve in a liquid's surface created by surface tension from the sides of its container

mesoderm — middle primary embryonic germ layer which lines the coelom and gives rise to muscle tissue, skeletal tissue, reproductive tissue and most circulatory tissue

mesoglea — acellular, gelatinous substance which fills the space between the outer and inner tissue layers of cnidarians; responsible for the "jelly-like" feel of these organisms

mesolecithal — classification of eggs containing a moderate amount of yolk

metamorphic — characteristic of having different body forms during the lifecycle

micronucleus — organelle of ciliates that represents a typical eukaryotic nucleus that contains the organism's entire genome and is responsible for genetic recombination

micropipettors — devices used to precisely aliquot minute volumes of fluid, generally less than 1 milliliter

microscopic protists — informal designation for the collection of protists that are generally too small to be observed without magnification

mitosis — the process of nuclear division in eukaryotes whereby two genetically identical daughter cells are produced from one event of cell division

molting — (syn. = ecdysis) the periodic act of shedding the skin or exoskeleton

monoecious — referring to an organism that contains both male and female reproductive structures

monophyletic — pertaining to a taxon derived from a single ancestral species that gave rise to no species in any other taxa

morphogenesis — development of body shape

morula — embryonic stage of development characterized by 16–32 cells

motile — having or pertaining to the ability to move

mouth — external opening to the digestive tract

muscle belly — thicker, middle section of muscle containing the bulk of muscle fibers

muscle tissues — groups of cells which permit movement of an animal through its environment and/or movement of substances through the animal

myelin — proteinaceous substance which coats the sheaths of the axons of nerve cells

myofibrils — composites of many individual muscle cells, giving these fibers a multi-nucleated appearance

myomeres — segmented muscular bundles that provide movements for swimming in many aquatic chordates

myosin — thick, proteinaceous filaments that overlap and alternate with actin filaments to produce muscular contractions

myxamoeba — individual amoeboid cell that, when in large numbers, makes up a cellular slime mold

-N-

nephridia — excretory organs in many invertebrates and some chordates (i.e., lancelet)

nerve cord — highly complex bundle of nerve fibers which, in many invertebrates, handles the majority of nervous coordination without intervention of the brain

nervous tissues — groups of cells which initiate and transmit electrical nerve impulses to and from the body parts and store information in the form of biochemical compounds

neural folds — thickening of ectodermal cells along the mid-dorsal region of the neurula to form two enlarged ridges on the surface

neural groove — depression along the mid-dorsal region of the neurula bordered by the neural folds

neural tube — enclosed cylinder which results from the meeting and fusing of the neural folds in the neurula; will develop into the brain and spinal cord

neuron — nerve cell composed of a cell body, axons and dendrites which transmits electrochemical information through the body

neurula — early embryonic stage of development characterized by the presence of 16 or 32 cells

neurulation — process of neural tube formation during embryonic development

nosepiece — revolving housing that supports microscope objective lenses

nostrils — (syn. = external nares) paired openings leading to olfactory receptors

notochord — cylindrical section of differentiated mesodermal cells below the neural tube in the neurula; anchors myomeres to provide tension for muscular contractions generating body movements

nucleus — membrane-bound organelle of eukaryotic cell containing the organism's DNA

-O-

objective lenses — microscope lenses of different magnification that work in conjunction with ocular lenses to magnify the image

ocular lens — microscope lens or lenses nearest the eye through which you look

oogamous — characteristic of gametes from opposite sexes that visibly differ in size, shape and often motility (i.e., sperm and eggs)

oogenesis — meiotic production of eggs (or ova)

oogonia — female germ cells

ootid — haploid cell receiving the bulk of the cytoplasm after the division (meiosis II) of a secondary oocyte

open circulatory system — system in which the blood is not confined within a network of vessels

opercula — paired bony plates which cover the gills of bony fish on either side of the head and allow for the release of water passing over the gills

oral — referring to the mouth

oral arms — cnidarian tentacles used for defense and prey capture

oral cirri — cephalochordate structures which collectively act as a strainer to exclude large particles from entering the mouth during filter feeding

oral disc — in anthozoans, the raised portion of the mouth (equivalent to the hypostome of hydrozoans)

oral groove — depression in some unicellular protists into which food is swept by ciliary currents

organogenesis — differentiation of organ tissues

orientation behavior — overt response that involves positioning or movements by an animal to a stimulus

origin — fixed or less-moveable end of a muscle attachment

osculum — large opening located at the top of sponges through which water which has collected in the spongocoel is pushed out

ossicles — calcium carbonate structures embedded beneath the epidermis of echinoderms and used for skeletal support

osteocytes — bone producing cells

ostia — (sing. = ostium) (1) openings in the cranial ends of the oviducts through which eggs released from the ovaries into the coelomic cavity enter the oviducts; (2) pores on the body surface of sponges through which water enters; (3) pores or openings that allow circulation of fluids between adjacent body or organ sections in sea anemones

ovary — female gonad that produces eggs

oviduct — tube that transports eggs from the ovaries to the uterus; in some vertebrates it secretes a gelatinous or calcified covering over the eggs for protection

ovipositor — external reproductive opening in found females of some arthropod species; bordered by pointy, chitinous teeth that penetrate the soil and create burrows for egg deposition

-P-

pancreas — granular organ usually located in close association with the duodenum and the stomach; produces digestive enzymes and a variety of hormones

parapodia — paired, feathery extensions along the lateral margins of polychaetes used primarily for respiration and, to some degree, for locomotion

parfocal — condition in which focal plane of microscope does not change substantially when switching between different objective lenses

pedal disc — tough, fleshy base that attaches cnidarian polyps to rocky substrates or sandy ocean floor

pedicellariae — pincer-like structures of echinoderms believed to kill small organisms that might settle on body surfaces, thus keeping the epidermis free of parasites and algae

pellicle — stiff outer covering of some unicellular protists that maintains basic cellular shape

penis — external male reproductive organ; deposits semen in the reproductive tract of the female and (in some animals) carries excretory wastes in the form of urine out of the body through the urethra

pennate — describes diatoms that are rod-shaped and have bilateral symmetry

perisarc — outer epidermal covering forming a raised sheath surrounding the coenosarc in some cnidarians

peristome — circular oral membrane in sea urchins which allows for eversion of teeth for feeding; opening at one end of *Vorticella* that serves as a mouth and is encircled by a ring of cilia

perpendicular grooves — characteristic pattern of grooves in dinoflagellates, in which their two flagella lay

phagocytosis — a type of endocytosis in which large food particles are engulfed

pharynx — muscularized portion of digestive tract usually responsible for pulling food into the digestive system

phycobilin — a class of accessory pigments that red algae possess; contains the pigments phycoerythrin (red) and phycocyanin (blue)

phycocyanin — a blue accessory pigment found in red algae

phycoerythrin — a red accessory pigment found in red algae

placenta — vascularized organ in mammals that unites fetus to the uterus during pregnancy and mediates exchange of nutrients and gases; absent in monotremes and marsupials

plankton — mostly microscopic organisms that drift passively or swim weakly near the surface of ponds, lakes and oceans

planula larva — swimming juvenile hydrozoan stage which settles to the bottom of the ocean floor where it develops into a new polyp

plasma — fluid matrix of blood

plasma membrane — the membrane at the boundary of a cell that acts as a selective barrier regulating the cell's chemical composition

plasmodial slime mold — type of macroscopic protist consisting of an acellular, wall-less mass of cytoplasm with numerous diploid nuclei

plasmodium — the vegetative (feeding) body form of a plasmodial slime mold

plastron — hard, bony, ventral portion of the shell of turtles

platelets — (syn. = thrombocytes) small, non-nucleated, colorless cell fragments present in mammalian blood which facilitate clotting

polar body — haploid cell receiving virtually no cytoplasm after the division (meiosis I) of a primary oocyte

polyp — stage in the cnidarian lifecycle represented by a cylindrical organism which remains attached to the substrate by a short stalk

polyphyletic — pertaining to a taxon whose members were derived from two or more ancestral forms not common to all members

power switch — control which turns microscope light on or off and may contain a rheostat (dimmer switch) that permits further adjustment of light intensity

preputial orifice — external opening of the penis in male vertebrates

primary oocytes — diploid cells resulting from the mitotic division of oogonia

primary spermatocytes — diploid cells resulting from the mitotic division of spermatogonia

proctodeum — posterior-most chamber of the cloaca in reptiles and birds; acts as a general collecting area for digestive and excretory wastes

proglottids — serially repeated "segments" of a tapeworm's body containing primarily reproductive organs; each can live for a limited time after it detaches from the main body

prosopyles — tiny pores scattered along the folds of the incurrent canals of sponges

prostomium — fleshy extension of the first body segment in oligochaetes that partially covers the opening to the mouth

prostate gland — accessory male reproductive gland that contributes fluid to the semen

protostome — animal whose mouth develops from the first embryonic opening and whose embryonic cells divide by spiral cleavage

proventriculus — digestive passageway in birds between the crop and gizzard that mixes peptic enzymes into the food

proximal — referring to a point of reference nearer the median plane or point of attachment on the body than another structure

proximate cause — series of immediate physiological events that lead to a behavior

pseudocoelom — a body cavity that lies between a layer of mesoderm and a layer of gastrodermis

pseudocoelomate — animal with a central body cavity that lies between gastrodermis and mesoderm

pseudopodia — cytoplasm-filled extensions of the plasma membrane employed by amoeboid species for locomotion and feeding

pygostyle — the caudal portion of the vertebral column in birds formed by the fusion of several lumbar vertebrae

pyloric chamber — second stomach region in some arthropods and echinoderms where chemical digestion of food occurs

pyrenoid — organelle found in autotrophic protists for starch formation and storage

-R-

radial canals — (1) flagellated chambers into which water is channeled from the prosopyles in the sponge; (2) portions of the echinoderm water vascular system emanating from the ring canal

radial symmetry — body parts arranged around a central axis such that any plane passing through the central axis divides the body into two similar halves

rectum — distal end of the intestinal tract; primary function is to reabsorb water and produce dry, concentrated feces

red tides — ecological phenomenon caused by massive population explosions of dinoflagellates

regeneration — biological process of regrowing and reshaping tissues into exact replicas of missing body parts

renal portal system — circulatory pathway in reptiles that channels all blood from the lower half of the body to the kidneys before sending it through the rest of the systemic circulatory system

respiratory trees — feathery, gill-like structures in sea cucumbers for gas exchange

Rhodophyta — taxonomic name for the monophyletic group containing all red algae

ring canal — portion of the echinoderm water vascular system encircling the mouth

rostral — referring to a point of reference closer to the tip of the nose

-S-

sagittal plane — longitudinal section separating the animal into right and left sides

salivary glands — special glands located within the oral cavity and neck that produce a variety of fluids and enzymes that facilitate swallowing and digestion

salt glands — osmoregulatory organs embedded in the orbits of the eyes in birds; secrete excess sodium chloride

sarcomeres — fundamental repeating units of striated muscle

scales — epidermally-derived, flattened plates forming part of the external body covering of an animal

sclerotium — irregular, resistant, hardened mass of a plasmodial slime mold that acts as a resting state during unfavorable environmental conditions

scolex — cranial end of tapeworm; lacks sensory structures but possesses modifications for attachment to intestinal wall of host

scrotum — pouch extending from the caudal region of some male mammals which contains the testes (after they have descended from the abdominal cavity during embryonic development); its presence allows the temperature of the testes to be maintained at a slighter lower temperature than that of the abdominal cavity

scutes — enlarged scales present in turtles

secondary oocyte — haploid cell receiving the bulk of the cytoplasm after the division (meiosis I) of a primary oocyte

secondary spermatocyte — haploid cell resulting from the division (meiosis I) of primary a spermatocyte

segmentation — the repetition of animal body regions containing similar organs

seminal vesicles — accessory male reproductive glands that contribute fluid to the semen

seminiferous tubules — tubule system located inside the testes in which sperm are produced through meiosis; primary spermatocytes are formed along the outer margins of the seminiferous tubules and migrate inward as they mature

serological pipettes — devices containing incremental markings for precisely aliquotting small volumes of fluid, generally between 1-20 milliliters

setae — small, hairlike bristles used for locomotion in annelids

sexual reproduction — type of reproduction in which offspring with unique genetic combinations are created by the union of sets of genetic material, usually, but not necessarily, from two separate parents

shell — hard, exterior secretion produced by the mantle of molluscs for protection

simple columnar epithelium — single layer of columnar cells in a tissue covering or lining

simple cuboidal epithelium — single layer of cuboidal cells in a tissue covering or lining

simple squamous epithelium — single layer of flattened cells in a tissue covering or lining

siphon — hollow tube found in cephalopods through which water is expelled from the mantle cavity at high velocity to propel the animal through the water

siphon retractor muscles — long muscles which control the contraction of the siphon in cephalopods

skeletal muscle — striated muscle tissue which consists of long, unbranched myofibrils which have a multinucleated appearance

small intestine — region of the digestive system responsible for nutrient absorption; consists of the duodenum, the jejunum and the ileum

smooth muscle — non-striated muscle tissue that consists of long, spindle-shaped fibers containing single nuclei

spectacle — clear scale covering the eyes of snakes

spermatic cord — long, narrow tube that leads from the testis through the abdominal wall and contains the vas deferens, the spermatic artery and vein, lymphatic vessels and numerous nerves

spermatid — haploid cell resulting from the division (meiosis II) of a secondary spermatocytes

spermatogenesis — the meiotic production of sperm cells

spermatogonia — male germ cells

spermatozoa — haploid, differentiated spermatids, or mature sperm cells

spicules — (1) hard, crystalline calcium carbonate or silicon structures which form the skeleton in many sponges; (2) spiny projections found on the tail of male nematodes that are used during copulation

spindle fibers — proteinaceous matrix of microtubules that form between the two pairs of centrioles during prophase

spines — calcareous projections in echinoderms that afford protection and support and are used for locomotion in some groups

spiracles — external openings in abdomen of insects that allow air flow into and out of tracheae

spleen — ductless, vascular organ in the abdominal cavity that is a component of the circulatory system; stores blood, recycles worn-out red blood cells and produces lymphocytes

spongin — proteinaceous, flexible material secreted by some sponges to form the skeleton

spongocoel — large central cavity that passes through the center of sponges

sporangiophore — stalk and spore-producing capsule of slime molds and fungi

stage adjustment knobs – controls which move microscope stage to center slide under objective lens

stage clips – attachments that hold slide in steady, stationary position on microscope stage

stage – platform that supports slides on microscope for viewing

starch – storage polysaccharide consisting of long polymers of glucose molecules; found in plants and some protists

stipe – thin, elongated region of algae modified to support the photosynthetic blades of the thallus

stomach – digestive chamber that typically stores food and assists in mechanical and chemical breakdown of the food

stone canal – portion of echinoderm water vascular system leading from the madreporite to the ring canal

Stramenopila – monophyletic candidate kingdom of protists containing two flagella of unequal length, the longer of which bears rows of tubular hairs

stratified squamous epithelium – several layers of flattened cells in a tissue covering or lining which collectively serve as a barrier against foreign substances and injury

suckers – specialized structures common in trematodes and cestodes for attachment of the parasite to its host

synsacrum – skeletal structure formed by the fusion of the 13 fused caudal vertebrae in birds

syrinx – structure at the caudal end of the trachea in birds that produces a wide array of calls and vocalizations

systemic heart – large, muscularized chamber in cephalopods that receives oxygenated blood from the gills and pumps it throughout the body

-T-

tail – projection from caudal end of body; in chordates the tail always contains some portion of the skeletal support system (i.e., notochord or vertebrae)

taxis – directed behavior in which an animal orients or moves toward or away from a stimulus

tentacles – long, extensible, prehensile found in cnidarians and some molluscs which aid in defense, prey capture and locomotion

test – support structure in sea urchins composed of numerous calcareous plates located beneath the epidermis that form the endoskeleton

testis – male reproductive organ that produces sperm

tetrads – homologous pairs of replicated double-stranded chromosomes joined together along the equator of the cell during meiosis I

thallus – name for the leaf-like body of many algae and lower, non-vascular plants

trachea – (pl. = tracheae) (1) cartilaginous tube extending from oral cavity to lungs through which air is transported during respiration; (2) one of many respiratory tubules of the grasshopper that conducts air flow from the outside environment directly to the tissues within the body

transverse plane – section perpendicular to the long axis of the body separating the animal into cranial and caudal portions

triploblastic – referring to an animal whose body develops from three embryonic tissue layers

tube feet – small, extensible protrusions connected to the water vascular system that permit locomotion and prey capture in echinoderms

-U-

ultimate cause – resultant evolutionary advantages that have promoted a particular behavior to remain in the animal's repertoire of possible responses

undulating membrane – region of trypanosome consisting of a fold in the plasma membrane and a single flagellum which together propel the animal through the host's bloodstream

unicellular – consisting of or having one cell

uniramous – unbranched

ureter – tube that transports urine from the kidney to the urinary bladder (in most animals) for storage

urethra – tube that leads from the urinary bladder to the outside of the body; transports urine and (in males) semen

urinary bladder – membranous sac that serves as a receptacle for filtrate from the kidneys

urodeum – middle chamber of the cloaca in reptiles and birds; receives reproductive and urinary products

urogenital opening – external opening for the release of urine and reproductive products

uterine horns – branched extensions of the body of the uterus; often the site of implantation and embryonic development in many mammals

uterus – region of female reproductive tract in which embryonic development of eggs or fetuses occurs

-V-

vagina – female reproductive chamber which receives penis during copulation; also serves as part of the birth canal

vasa efferentia – small ducts in male amphibians that transport sperm from the testes to the kidneys

vas deferens – (syn. = ductus deferens) tubule in male animals that transports sperm

vector – an organism that acts as a secondary host to a parasite and transmits the parasite to its primary host, usually without suffering any ill effects itself

vegetal pole – portion of an egg toward which the yolk is segregated and which appears lighter in coloration in comparison to the animal pole

veins – vessels which conduct blood toward the heart

ventral – referring to a point of reference nearer the underside of the body

ventricle – muscularized chamber of the heart which forcefully contracts and expels blood into the arterial system

vertebrae – bony segments which constitute the spinal column in many animals

visceral mass – soft, fleshy pouch containing the major internal organs of molluscs

viviparous – referring to an animal which produces live young instead of eggs

vomeronasal organ – (syn. = Jacobson's organ) chemosensory organ located in the roof of a snake's mouth into which the forks of the tongue are placed, thereby depositing odor molecules for detection and interpretation by the brain

vulva – most caudal region of the female urogenital tract consisting of the vestibule, clitoris and labia

-W-

water vascular system – hydraulic system composed of a network of canals and small pumps that move water through the body of echinoderms to affect locomotor movements

wheel organ – cephalochordate structure in oral cavity; lined with cilia to produce water current that brings food into the mouth

wings – modified appendages used for flight

working distance – space between the objective lens and the slide when using a microscope

-Y-

yolk mass – nutrient source for the developing embryo in egg-laying animals

yolk plug – small remnant of nutritive source present at the opening of the blastopore during embryonic development in certain groups

-Z-

zygote – diploid product of the fusion of two haploid gametes during fertilization